SPECTRUM MANAGEMENT FOR SCIENCE IN THE 21st CENTURY

Committee on Scientific Use of the Radio Spectrum

Committee on Radio Frequencies

Board on Physics and Astronomy

Division on Engineering and Physical Sciences

NATIONAL RESEARCH COUNCIL
OF THE NATIONAL ACADEMIES

THE NATIONAL ACADEMIES PRESS
Washington, D.C.
www.nap.edu

THE NATIONAL ACADEMIES PRESS 500 Fifth Street, N.W. Washington, DC 20001

NOTICE: The project that is the subject of this report was approved by the Governing Board of the National Research Council, whose members are drawn from the councils of the National Academy of Sciences, the National Academy of Engineering, and the Institute of Medicine. The members of the committee responsible for the report were chosen for their special competences and with regard for appropriate balance.

This study was supported by the National Science Foundation under Grant No. 0410006, by the National Aeronautics and Space Administration under Award No. NNH05CC15C, and by the Department of Commerce under Award No. DG133R04CQ0009, TO #26. Any opinions, findings, conclusions, or recommendations expressed in this publication are those of the author(s) and do not necessarily reflect the views of the organizations or agencies that provided support for the project.

International Standard Book Number-13: 978-0-309-14686-9
International Standard Book Number-10: 0-309-14686-0

Copies of this report are available from the National Academies Press, 500 Fifth Street, N.W., Lockbox 285, Washington, DC 20055; (800) 624-6242 or (202) 334-3313 (in the Washington metropolitan area), Internet: http://www.nap.edu; and from the Board on Physics and Astronomy, National Research Council, 500 Fifth Street, N.W., Washington, DC 20001, Internet: http://www.national-academies.org/bpa.

Printed in the United States of America

THE NATIONAL ACADEMIES

Advisers to the Nation on Science, Engineering, and Medicine

The **National Academy of Sciences** is a private, nonprofit, self-perpetuating society of distinguished scholars engaged in scientific and engineering research, dedicated to the furtherance of science and technology and to their use for the general welfare. Upon the authority of the charter granted to it by the Congress in 1863, the Academy has a mandate that requires it to advise the federal government on scientific and technical matters. Dr. Ralph J. Cicerone is president of the National Academy of Sciences.

The **National Academy of Engineering** was established in 1964, under the charter of the National Academy of Sciences, as a parallel organization of outstanding engineers. It is autonomous in its administration and in the selection of its members, sharing with the National Academy of Sciences the responsibility for advising the federal government. The National Academy of Engineering also sponsors engineering programs aimed at meeting national needs, encourages education and research, and recognizes the superior achievements of engineers. Dr. Charles M. Vest is president of the National Academy of Engineering.

The **Institute of Medicine** was established in 1970 by the National Academy of Sciences to secure the services of eminent members of appropriate professions in the examination of policy matters pertaining to the health of the public. The Institute acts under the responsibility given to the National Academy of Sciences by its congressional charter to be an adviser to the federal government and, upon its own initiative, to identify issues of medical care, research, and education. Dr. Harvey V. Fineberg is president of the Institute of Medicine.

The **National Research Council** was organized by the National Academy of Sciences in 1916 to associate the broad community of science and technology with the Academy's purposes of furthering knowledge and advising the federal government. Functioning in accordance with general policies determined by the Academy, the Council has become the principal operating agency of both the National Academy of Sciences and the National Academy of Engineering in providing services to the government, the public, and the scientific and engineering communities. The Council is administered jointly by both Academies and the Institute of Medicine. Dr. Ralph J. Cicerone and Dr. Charles M. Vest are chair and vice chair, respectively, of the National Research Council.

www.national-academies.org

COMMITTEE ON SCIENTIFIC USE OF THE RADIO SPECTRUM

MARSHALL H. COHEN, California Institute of Technology, *Co-Chair*
ALBIN J. GASIEWSKI, University of Colorado at Boulder, *Co-Chair*
DONALD C. BACKER, University of California, Berkeley
ROBERTA BALSTAD, Columbia University
STEVEN W. ELLINGSON, Virginia Polytechnic Institute and State University
DARREL EMERSON, National Radio Astronomy Observatory
AARON S. EVANS, University of Virginia and National Radio Astronomy Observatory
JOEL T. JOHNSON, Ohio State University
PAUL KOLODZY, Kolodzy Consulting, LLC
DAVID B. KUNKEE, The Aerospace Corporation
MOLLY K. MACAULEY, Resources for the Future, Inc.
JAMES M. MORAN, Harvard-Smithsonian Center for Astrophysics
LEE G. MUNDY, University of Maryland at College Park
TIMOTHY J. PEARSON, California Institute of Technology
CHRISTOPHER S. RUF, University of Michigan
FREDERICK S. SOLHEIM, Radiometrics Corporation
DAVID H. STAELIN, Massachusetts Institute of Technology
ALAN B. TANNER, Jet Propulsion Laboratory

Staff

DONALD C. SHAPERO, Director, Board on Physics and Astronomy
BRIAN D. DEWHURST, Program Officer (through July 2009)
DAVID LANG, Program Officer
MERCEDES ILAGAN, Administrative Assistant (through February 2008)
CARYN KNUTSEN, Program Associate (from March 2008)
VAN AN, Financial Associate (through May 2008)
BETH DOLAN, Financial Associate (from June 2008)

COMMITTEE ON RADIO FREQUENCIES

PAUL A. VANDEN BOUT, National Radio Astronomy Observatory, *Chair*
JEFFREY PIEPMEIER, NASA Goddard Space Flight Center, *Vice-Chair*
ANA P. BARROS, Duke University
DOUGLAS C.-J. BOCK, University of California, Berkeley/Combined Array for Research in Millimeter-wave Astronomy
STEVEN W. ELLINGSON, Virginia Polytechnic Institute and State University
DAVID G. LONG, Brigham Young University
DARREN McKAGUE, University of Michigan
JAMES M. MORAN, Harvard-Smithsonian Center for Astrophysics
MELINDA PIKET-MAY, University of Colorado at Boulder
STEVEN C. REISING, Colorado State University
ALAN E.E. ROGERS, Massachusetts Institute of Technology/Haystack Observatory
LUCY ZIURYS, University of Arizona

Staff

DONALD C. SHAPERO, Director, Board on Physics and Astronomy
DAVID B. LANG, Program Officer
CARYN J. KNUTSEN, Research Associate

Preface

In the early years of the 21st century, policy officials recognized both the need for additional blocks of frequencies in the electromagnetic spectrum for new technologies and the desires of existing users to obtain additional bandwidth. A number of activities were thus begun, with the goals of identifying unused frequencies and suggesting methods by which the regulatory structure could encourage their more efficient use. In June 2002, the Federal Communications Commission (FCC) formed the Spectrum Policy Task Force for the following purposes:

1. To provide specific recommendations to the FCC for ways in which to evolve the current "command-and-control" approach to spectrum policy into a more integrated, market-oriented approach that provides greater regulatory certainty while minimizing regulatory intervention; and
2. To assist the FCC in addressing ubiquitous spectrum issues, including interference protection, spectral efficiency, effective public-safety communications, and implications of international spectrum policies.

The Spectrum Policy Task Force concluded that "while the commission has recently made some major strides in how spectrum is allocated and assigned in some bands, principally through flexible rules and competitive bidding, spectrum policy is not keeping pace with the relentless spectrum demands of the market. The task force has begun the process of reexamining 90 years of spectrum policy to ensure that the commission's policies evolve with the consumer-driven evolution of new wireless technologies, devices, and services."[1]

Recognizing the growing importance of radio observations to their respective missions and the increasing potential for interference from new wireless technologies, NASA, the Department of Commerce, and the National Science Foundation commissioned the National Research Council (NRC) to identify the spectrum needs of today's scientific activities and to assist spectrum managers in balancing the requirements of scientific uses of the spectrum with those of other interests. This report is written in response to that request. The committee discussed its original charge at length and chose to consider only the passive ("receive-only") scientific applications of the radio spectrum, and specifically how the requirements for spectrum could be expected to evolve over the next two decades.[2] This decision did not imply any prioritization of the active versus passive scientific uses of the spectrum, but instead stemmed from the committee's recognition that passive scientific uses involve unique issues and that the committee had a limited amount of time in which to complete its task.

To address its task, the NRC's Committee on Scientific Use of the Radio Spectrum—comprising representatives of universities, private industry, and non-profit organizations[3]—employed four in-person meetings, four town hall meetings, and numerous teleconferences in the development of its report. The committee's work was aided by presentations from a number of outside experts who provided detailed information at in-person meetings.

The committee focused on three major topics: Earth remote sensing (see Chapter 2), radio astronomy (see Chapter 3), and interference mitigation (see Chapter 4). It conducted an in-depth study of each of the major topics, including the current and expected future status of Earth remote sensing and radio astronomy and applicable radio frequency interference mitigation technologies. The committee developed a series of findings on the basis of the material presented in these chapters, together with an associated series of recommendations, to help ensure the viability of these scientific endeavors. The findings and recommendations are detailed in Chapter 5. As dictated by the statement of task, the committee did not make recommendations on the allocation of specific frequencies, but it did comment on spectrum use by the relevant scientific communities and how it might be protected in the future.

This report attempts to lay the foundation of an effort to identify the spectrum needs of radio astronomy and Earth remote sensing, identify the benefits of these two activities, and develop practical, forward-looking approaches to spectrum

---

[1]Federal Communications Commission, *Report of the Spectrum Policy Task Force*, Washington, D.C., November 2002.

[2]The committee's statement of task appears in Appendix A.

[3]Biosketches of the members of the committee are provided in Appendix B.

access that are needed to ensure the necessary conditions for their important observations.

It is noted that a report on the uses of passive service[4] bands for both Earth remote sensing and radio astronomy by a panel of the NRC's Committee on Radio Frequencies (CORF) was published in 2007.[5] The present report differs from the 2007 report in assessing both the current and future uses of the passive services. This report also includes a focus on technology for interference mitigation.

[4]The passive services are those for which the signal is produced by nature and the applications are "receive-only."

[5]National Research Council, *Handbook of Frequency Allocations and Spectrum Protection for Scientific Uses*, Washington, D.C.: The National Academies Press, 2007.

Acknowledgment of Reviewers

This report has been reviewed in draft form by individuals chosen for their diverse perspectives and technical expertise, in accordance with procedures approved by the National Research Council's (NRC's) Report Review Committee. The purpose of this independent review is to provide candid and critical comments that will assist the institution in making its published report as sound as possible and to ensure that the report meets institutional standards for objectivity, evidence, and responsiveness to the study charge. The review comments and draft manuscript remain confidential to protect the integrity of the deliberative process. We wish to thank the following individuals for their review of this report:

Paul Feldman, Fletcher, Heald & Hildreth, PLC,
Dale N. Hatfield, Independent Consultant, Longmont, Colorado
Anthony Janetos, Pacific Northwest National Laboratory,
Roger Lang, George Washington University,
Michael Marcus, Marcus Spectrum Solutions,
Thomas Meissner, Remote Sensing Systems, Inc.,
Steven Reising, Colorado State University,
Chris Salter, National Astronomy and Ionosphere Center, Cornell University,
Paul Vanden Bout, National Radio Astronomy Observatory,
William "Jack" Welch, University of California, Berkeley, and
David Woody, California Institute of Technology, Owens Valley Radio Observatory.

Although the reviewers listed above have provided many constructive comments and suggestions, they were not asked to endorse the conclusions or recommendations, nor did they see the final draft of the report before its release. The review of this report was overseen by Martha Haynes, Cornell University. Appointed by the NRC, she was responsible for making certain that an independent examination of this report was carried out in accordance with institutional procedures and that all review comments were carefully considered. Responsibility for the final content of this report rests entirely with the authoring committee and the institution.

Contents

APPENDIXES

Summary

Natural radio emissions from objects as diverse as hurricanes and distant galaxies yield vital information about the planet Earth and its place in the universe. Observations of Earth are central to weather forecasting and climate studies, and radio observations of the cosmos are similarly critical for increasing our understanding of the universe and answering grand questions such as that on the origin of planets. Such information is gathered by geoscientists using complex Earth-orbiting satellites and ground-based equipment and by radio astronomers using large, ground-based radio telescopes. Signals from natural radio emissions are extremely weak, and the equipment used to measure them is becoming more and more sophisticated and sensitive.

The radio spectrum is also being used by radiating, or "active," services, ranging from aircraft radar to rapidly expanding consumer services such as cellular telephony and wireless Internet. These valuable active services transmit radio waves and thereby potentially interfere with the receive-only, or "passive," scientific services, which do not radiate. Transmitters for the active services create an artificial "electronic fog," which can cause confusion and, in severe cases, totally blind the Earth Exploration-Satellite Service (EESS) and Radio Astronomy Service (RAS) receivers.

Both the active and the passive services are increasing their use of the spectrum, and so the potential for interference, already strong, is also increasing. This tension between the active services' demand for greater spectrum use and the passive users' need for quiet spectrum is at the heart of this report's discussion and motivates the findings and recommendations of the study committee—the National Research Council's Committee on Scientific Use of the Radio Spectrum.

Many billions of dollars have been invested in the nation's radio astronomy and Earth remote sensing facilities. The public marvels at new discoveries made at radio astronomy observatories, and the nation remains ever more reliant on accurate and up-to-date weather and climate information retrieved from Earth remote sensing satellites. The use of the radio spectrum to obtain these observations is regulated and protected in accordance with national and international spectrum rules, but the relatively recent proliferation of wireless technology is challenging engineers' abilities to mitigate unwanted interference from the active services.

Complex rules govern the occupancy and use of the electromagnetic spectrum, both nationally and globally, but these rules have not adequately evolved with technology. Inefficiencies in spectrum use exist while demand increases, and most regulations are not aligned with or even cognizant of the special needs of passive scientific users. These issues are identified in this report, and addressing them presents the nation with an exceptional opportunity to adapt to the wireless revolution while protecting the passive users of the radio spectrum.

The radio spectrum is a finite resource that has been managed as such for the past 70 years by the federal government. This management enabled the growth of strong commercial and scientific communities. The endless pursuit of better techniques to leverage the unique characteristics of the radio spectrum has led to discoveries and innovations of enormous scientific and societal value. Over the past 20 years, rapid technological improvements have exponentially increased the capabilities of scientific, commercial, and government users. But today, the current regulatory regime is straining to enable the capabilities and meet the needs of the various communities of users. A new path is needed to preserve access to the radio spectrum, in which important scientific discoveries are made and civilian and government remote sensing operations are conducted, while allowing for growth that serves an increasingly mobile society. This Summary presents the report's key findings and recommendations.

Finding: Passive remote sensing observations are essential for monitoring Earth's natural systems and are therefore critical to human safety, the day-to-day operations of the government and the private sector, and the policy-making processes governing many sectors of the U.S. economy.

Finding: Radio astronomy has great potential for further fundamental discoveries, including the origins and evolution of the universe, the nature of matter, and life in other solar systems, which will have an enormous impact on our understanding of fundamental physics and the place of humanity in the universe.

Recommendation: Recognizing that the national investment in passive radio astronomy and Earth remote sensing is dependent on access to the radio spectrum, the committee recommends that the Federal Communications Commission (FCC) and the National Telecommunications and Information Administration (NTIA) ensure that access to spectrum for passive radio and microwave observations of Earth environmental variables and radio astronomical observations of the sky is protected in the development of future spectrum policy.

Technological innovations continue to increase the utility of the radio spectrum. The advent of new technologies designed to exploit the diversity of the radio spectrum in space, frequency, polarization, and time will increase the efficiency of its use. However, the current means for managing spectrum use must be changed, as the current policies threaten to thwart scientific discovery, diminish the usefulness of critical environmental observations, and limit economic growth because of the inefficient use of finite spectral resources. Therefore, new spectrum management policies need to be explored for the sake of ensuring these critical national capabilities.

Finding: Radio wave bands (10 MHz to 3 THz) are indispensable for collecting information associated with specific astronomical and environmental phenomena. Often the same bands are equally indispensable for both passive Earth remote sensing and radio astronomy, and the passive nature of both services enables them to share the spectrum productively. Currently, 2.07 percent of the spectrum below 3 GHz is allocated to the RAS and EESS on a primary basis, and 4.08 percent is allocated on a secondary basis (measured in hertz).

Finding: Important scientific inquiry and applications enabled by the Earth Exploration-Satellite Service and the Radio Astronomy Service are significantly impeded or precluded by radio frequency interference (RFI). Such RFI has reduced the societal and scientific return of EESS and RAS observatories and necessitates costly interference mitigation, which is often insufficient to prevent damage from RFI.

Finding: Better utilization of the spectrum and reduced RFI for scientific as well as commercial applications are possible with better knowledge of actual spectrum usage. Progress toward these goals would be made by gathering more information through improved and continuous spectral monitoring. This would be beneficial to both the commercial and the scientific communities.

Recommendation: The Department of Commerce/National Telecommunications and Information Administration (NTIA), in collaboration with the National Science Foundation (NSF), NASA, and the National Oceanic and Atmospheric Administration (NOAA), should spearhead the development of a national spectrum assessment system that measures the radio frequency (RF) environment with appropriately high resolution in time, space, and frequency for purposes of spectrum development and management, based on the spectral and spatial density of emitters.

The next generation of spectrum management policies must enable better sharing of the spectrum as well as diminishing the impact that users have on the RF spectrum. This can be done by exploiting currently available technologies and hastening the development of nascent technologies. New policies should encourage the following:

- The development of the means for direct interaction between active and passive spectrum users to protect current and future scientific uses of the spectrum. The nation needs to provide the policies that will make the spectrum more useful and productive for all users.

Recommendation: The EESS and RAS communities should be provided additional support through NSF, NASA, and NOAA to increase their participation in spectrum management forums within the International Telecommunication Union (ITU), FCC, NTIA, and other organizations. The goal of such participation is to foster outreach, advance the understanding of interference and regulation issues, and initiate mutual cooperation for interference mitigation.

- The development and implementation of technology to address RFI for current and future satellite systems to ensure that the national investment in scientific uses of the spectrum is preserved.

Recommendation: Investment in the development of mitigation technology should be increased so that it is commensurate with the costs of data denial that result from the use of systems without mitigation. To this end, NSF and NASA should support research and development for unilateral[1] RFI mitigation technology in both EESS and RAS systems. NASA, NOAA, and the Department of Defense should require that appropriate RFI analyses and tests and practical RFI mitigation techniques be applied to all future satellite systems carrying passive microwave sensors.

- A regulatory environment that enables sharing the spectrum in both space and time. This is a "win-win" scenario that will enable additional scientific uses without impacting commercial development.

Recommendation: The NSF, NASA, and NTIA should jointly support research and development for cooperative RFI mitigation techniques and the associated forums

[1]As discussed in Chapter 4.

and outreach necessary to enable the development of standards for greater spectral utilization and interference avoidance.

Recommendation: As cooperative spectrum-sharing techniques come into use, NSF and NASA spectrum managers should work with the regulatory agencies to enable observations that require an extremely wide spectral range. Such observations would provide a useful metric for the effectiveness of spectrum-sharing techniques for the passive services.

These new initiatives are not easy, nor will they make success a certainty. It will take a national effort to understand clearly the needs of both communities, scientific and commercial, and to motivate each to make the choices necessary to enable greater access for each to the radio spectrum. The next generation of scientific users of the radio spectrum needs to be afforded the capacity to develop the technology to seek new horizons.

Recommendation: The Office of Science and Technology Policy should create a new, permanent, representative technology advisory body to identify technical and regulatory opportunities for improving spectrum sharing among all active and passive users, both government and nongovernment.

In one sense, spectrum used for passive purposes, including Earth remote sensing and radio astronomy, can be likened to parkland preserved for public use. The true societal value of small parcels of land, especially in crowded urban areas, defies monetization, and proactive measures are required to ensure the preservation and shared use of such land. A small fraction of the radio spectrum allocated for passive purposes performs a similarly valuable societal function and requires proactive management to remain available—in this case for scientific purposes. The passive services both offer a critical return to society through operations in support of environmental prediction and provide scientific intellectual value. Although the impacts of the passive services are difficult to quantify, they are valuable to society for providing vital information for climate and weather studies and in allowing astronomical studies of the heavens. The quiet radio bands, like public parks, deserve protection.

It would be in the strongest interests of the nation to ensure that access to spectrum for scientific purposes is maintained during the coming decades. The committee's recommendations provide a pathway for putting in place the regulatory mechanisms and associated supporting research activities necessary to accomplish this important task. The committee believes that such a pathway will also lead to greater efficiency in the active use of the spectrum, which should benefit all direct and indirect consumers of wireless telecommunications and data services.

1

Introduction

Natural radio emissions from objects as diverse as hurricanes and distant galaxies are employed by scientists to study the objects themselves. This information is vital for analyzing and forecasting weather and climate and for understanding the distant cosmos. The geophysical studies use remote sensing satellites and ground-based instruments, and the radio astronomy observations employ large, ground-based radio telescopes, or antennas. The techniques used by the two groups are fundamentally similar. The variations in the strength and polarization of the radio signals with direction, frequency, and time are measured with receivers of ever-increasing sensitivity and sophistication. The detailed techniques must be different because satellites pass quickly over any given spot on Earth, whereas a radio telescope can track a given source in the sky for hours. The two groups of users, working within the Earth Exploration-Satellite Service (EESS) and the Radio Astronomy Service (RAS), use many of the same frequency bands, and they have many interference problems in common.

The launch of the first U.S. Earth remote sensing satellite, TIROS-1, in 1960, ushered in an era of unprecedented scientific understanding of the planet as a complex system of systems. For the first time humanity was provided the opportunity to visualize and understand the interactions of many of Earth's constituent processes. This global view was only made possible by the development of advanced sensors that were able to take advantage of the new perspective offered by satellites. Chief among the new classes of sensors used in the nearly five decades since TIROS has been the passive microwave radiometer, which holds the unique advantage over optical and infrared systems of being able to probe through clouds. These sensors are passive in that they do not transmit any signals or communications; they only receive naturally occurring signals. Passive instruments are thus much different from active devices, with which people are most familiar. Active devices include cellular telephones, wireless Internet networks, garage-door openers—anything that emits a signal, whether purposefully or not. Owing to the unique purposes for which passive instruments are used, these instruments also have unique designs and needs, which are discussed in detail in this report. These passive devices are used by Earth scientists for economically and scientifically important observations of Earth's environment and by astronomers to observe the vast reaches of the cosmos beyond this planet.

The use of passive microwave sensors allows rainfall, clouds, ocean surface winds, temperature, and moisture distributions—the primary variables of meteorology—to be quantified over the globe, under clear and cloudy conditions and during both day and night. Data from passive microwave sensors are now a vital component in the complex calculations used for weather prediction.

In addition to enabling new understanding of atmospheric processes, passive microwave observations[1] have brought about a new understanding of Earth's surface processes. The distinct microwave signatures produced by water in its various phases (liquid, ice, vapor) permit all-weather measurements of environmental variables such as snowpack depth, soil moisture, sea ice extent, sea surface temperature, and sea surface salinity. These and other related passive microwave observables, including biomass and vegetation water content, are becoming increasingly important as drivers of industry and agriculture, particularly as global resources of freshwater, arable land, and fisheries are further stretched to satisfy an increasingly large and demanding global population. The role played by passive microwave observations from space as well as from surface-based and airborne platforms in the management of these resources and the understanding of their interactions with other natural systems cannot be overstated. Currently, 21 satellites carrying passive microwave sensors are orbiting Earth.

During roughly the same era in which Earth remote sensing was developed—but preceding it by about a decade—passive radio observations were used to study the makeup of the cosmos under what is now the discipline of radio astronomy. Radio astronomy is a young science, about 60 years old, but it has contributed enormously

[1]The terms *passive microwave observations, microwave brightness, microwave emission,* and similar terms that are used in Chapter 3 of this report, on the EESS (remote sensing), are synonymous with the terms *radio observations, radio brightness,* and *radio emission* that are used throughout Chapter 4, on radio astronomy. The popular use of *microwave* within the passive remote sensing community may have arisen in an attempt to distinguish microwave observations from visible and infrared remote sensing observations in which the Rayleigh-Jeans approximation is not applicable, and radiance power is expressed as an equivalent brightness temperature. The EESS passive microwave measurements are referenced to absolute temperature.

to the understanding of the universe. It provided the first view of the cosmos outside the optical band and revealed an extraordinary variety of remarkable objects and phenomena that had never been suspected, including pulsars (spinning condensed stars acting as rotating radio beacons), the cosmic microwave background radiation (showing the universe to have started in an initial explosion—the Big Bang—and to have been expanding ever since), gigantic molecular clouds where new stars are being born, and active galactic nuclei (in which reside giant black holes fed by a disk of gas and dust and from which emanates an enormously energetic pair of jets, going far across intergalactic space). In the near future, radio astronomical observations will provide insights about the events that occurred around the time that the first stars were born—known as the epoch of reionization.

Radio sciences have a strong practical importance. Accurate weather forecasting is vital for many activities critical to human health, welfare, and security—including agriculture, transportation, military defense, and the mitigation of damage from extreme weather events such as hurricanes, wildfire, and drought. These applications and others are discussed in detail in Chapter 2 of this report. The long-term monitoring of Earth is vital for climate assessment and prediction and is a major aid in the understanding of climate change—one of the most important problems currently facing humanity. The proper management of the environment both now and for decades to come will be contingent on remotely sensed data.

On a larger scale, radio astronomy opened people's view of the cosmos to take in the violent, enormously complex universe that it is now known to be. The shift in thought engendered by radio astronomy is analogous to that caused 500 years ago when it was first recognized that Earth orbits the Sun: the notion of Earth's special location in the universe disappeared, and people began seeing their world as an element of the cosmos rather than as its center. But radio astronomy also has practical applications in technology that support today's development of our information infrastructure, as discussed in Chapter 3. These include technical developments in high-performance antennas, sensitive radio receivers, electronics, computing, signal processing, and scientific education.

Unfortunately, human-made radio frequency interference (RFI) can make the radio science measurements more difficult, and in some cases it can render them unusable. This problem is introduced in §1.3 and §1.4 below and discussed in depth in Chapter 4.

1.1 THE PASSIVE RADIO SPECTRUM

Radio astronomy and passive Earth remote sensing both rely on detecting, recording, and interpreting weak natural radio frequency emissions. These emissions are radiated by all absorptive bodies: for example, forests, clouds, the Sun, and galaxies. The detailed features of the radiation provide information on the temperature, density, composition, motion, and other characteristics of the object or medium being observed. Earth studies and astronomical studies cover most of the radio spectrum, ranging from about 15 MHz (the lower limit for the radio transparency of Earth's ionosphere), to the current limits of radio technology at many hundreds of gigahertz. The highest radio frequencies (at 1 THz and above) merge with infrared radiation, and some studies require continuous measurements from the radio into the infrared bands, and even on to optical bands or beyond.

Natural radio emissions are generated by a variety of mechanisms. All matter emits "thermal noise," with a characteristic frequency spectrum that depends on its temperature. While hot objects, such as stars, emit mainly in the infrared, optical, and ultraviolet portions of the electromagnetic spectrum, cold gas, dust between the stars, and materials on Earth such as water, soil, and atmospheric gases with smaller temperatures (of a few hundred kelvin and below) emit mostly in the radio wave and submillimeter-wave portions of the spectrum (see Figure 1.1).

Radio radiation is also emitted from atoms and molecules when they move from one quantum state to another. This *line radiation* appears at characteristic

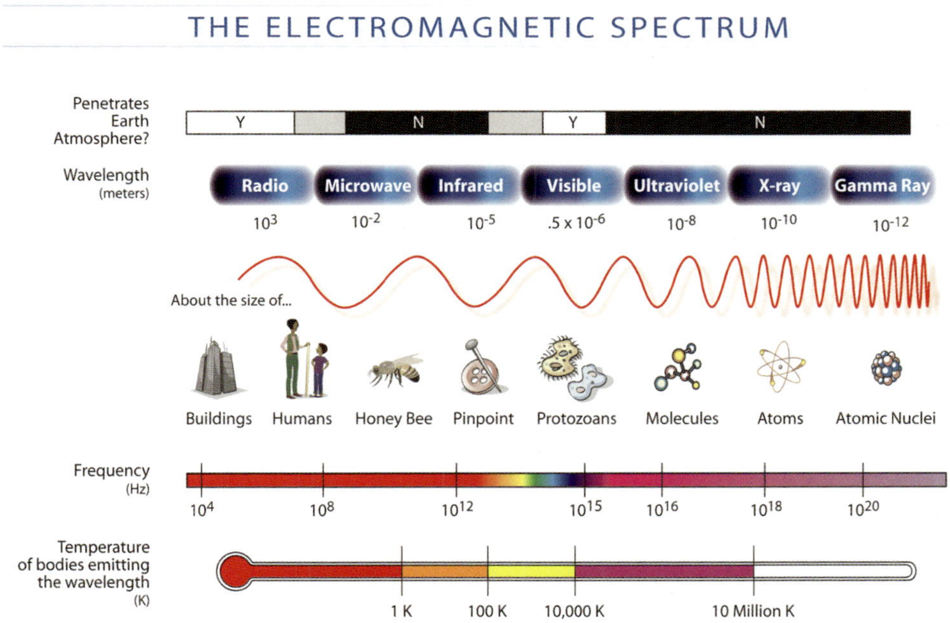

FIGURE 1.1 The characteristics of the electromagnetic spectrum. See Figures 1.2 and 1.3 in this chapter for a more detailed picture of the atmospheric penetration of electromagnetic waves. Image from NASA, *Science Program for NASA's Astronomy and Physics Division*, Washington, D.C., 2006.

frequencies determined by the particular quantum transition of the atomic or molecular species in question (Figures 1.2 and 1.3). The measurement of radiation at and near these transition frequencies is extremely important for both Earth science and radio astronomy. In Earth remote sensing, line radiation spectra can be used to obtain temperature and humidity profiles in the atmosphere from the surface of Earth on up into the mesosphere. Observations for such profiling purposes occur near the centers of atmospheric absorption lines and within the immediately adjacent "wings" on either side of the line centers. In radio astronomy, a proper interpretation of line radiation provides information on the composition, density, and temperature of the material under study. Radio astronomers are interested in frequency bands where an interesting atomic or molecular transition occurs and where Earth's atmosphere is particularly transparent. Spectra from these bands are often used to derive the motions in cosmic clouds, or in galaxies. Because different molecules—for example CO and HCN—have different excitation conditions, the study of several molecules (or several lines from one molecule) can give three-dimensional information about the cloud. This is analogous to the way that atmospheric profiles are found in Earth science measurements.

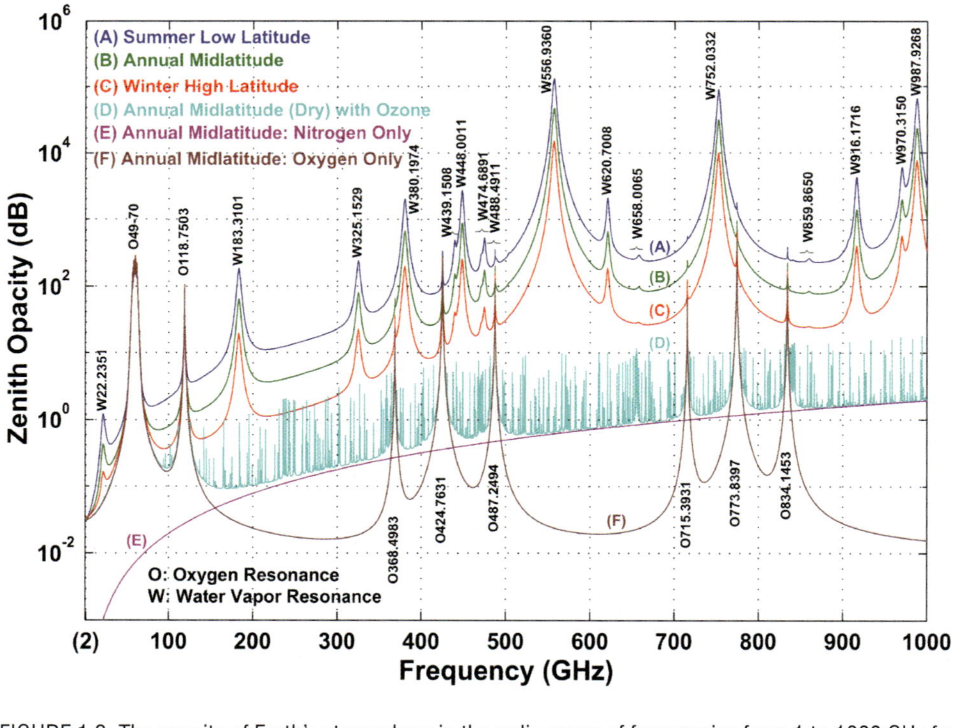

FIGURE 1.2 The opacity of Earth's atmosphere in the radio range of frequencies from 1 to 1000 GHz for six scenarios. Image courtesy of A.J. Gasiewski, University of Colorado.

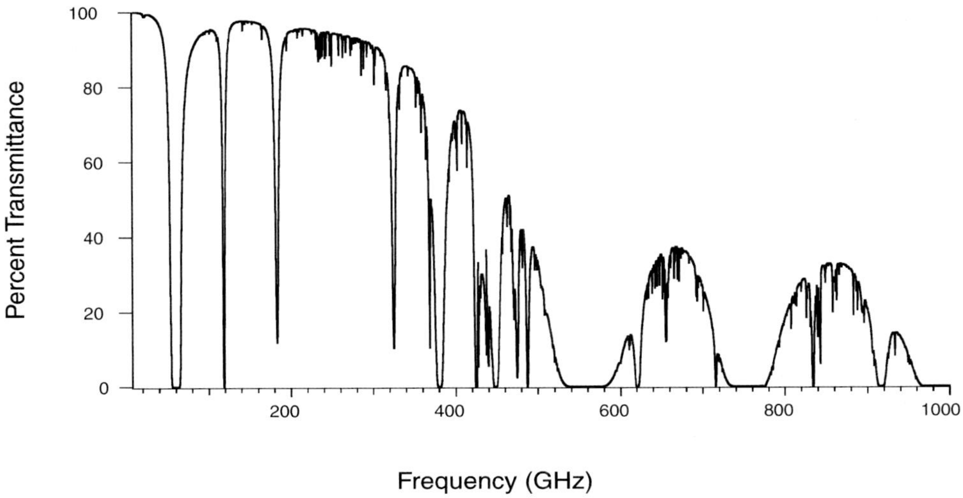

FIGURE 1.3 The transmission spectrum of Earth's atmosphere in the radio range of frequencies from 1 to 1000 GHz at Mauna Kea, Hawaii, a very dry site at an altitude of approximately 14,000 ft. Such high, dry sites are drier than Scenario E as given in Figure 1.2, making them suitable for astronomical observations above 200 GHz. Note that the water vapor line at 22 GHz (see Figure 1.2) causes negligible loss in transmission at this site, but the lines at 556 and 752 GHz are so strong—even on the high mountaintop—that the atmospheric transmission is essentially zero, and no astronomical observations can be made from 520 to 580 GHz and 730 to 780 GHz. Image courtesy of L. Ziurys, University of Arizona.

For many nongaseous materials on Earth (such as liquid water, ice, soil, snow, and vegetation), the radiation is broadened by the strong interaction of closely spaced molecules into a continuum that exhibits a slow spectral variation over a wide range of frequencies. Continuum radiation spectra can also occur when the scale of surface roughness or feature size (i.e., raindrop or cloud-particle diameter) is comparable to or smaller than a wavelength of the radiation. In Earth remote sensing, continuum radiation spectra are measured at a variety of microwave frequencies. Optimal frequencies for measuring continuum radiation are between the major transition frequencies for oxygen and water vapor (see Figures 1.2 and 1.3). In these "windows," the ability to probe through the atmosphere is maximized, thus making the continuum radiation easy to observe. Similar frequencies are used in radio astronomy for continuum measurements.

The Doppler effect, in which motion of the emitter toward or away from the observer shifts the received frequency, provides a means of determining the motion of the material. Doppler-shifted radiation enables the measurement of the rotation of matter in spiral galaxies and that of the motion of air in the upper atmosphere. The expansion of the universe leads to a similar shift in the frequencies of spectral lines that increase as the distance to the source increases. This effect, known as the cosmological redshift, allows distances to faraway galaxies to be accurately measured. (See Box 3.2, "Redshift," in Chapter 3 of this report.)

Another widespread emission mechanism is synchrotron radiation, generated by the acceleration of electrons in a magnetized plasma. Our Milky Way Galaxy is suffused with a dilute plasma that emits synchrotron radiation at frequencies of about 1 GHz and below. Over a much wider frequency range, this radiation is also associated with some of the most powerful events in the universe: pulsars, supernovae, gamma-ray bursts, and quasars, in which matter falling into a giant black hole radiates a prodigious amount of radiation; and radio galaxies, in which jets and giant cocoons of plasma ejected from a galaxy nucleus extend well outside the host galaxy. Synchrotron and thermal radiation are emitted across broad frequency bands and with a characteristic spectral shape. Their measurement is often not restricted to any one particular frequency, although when the band shape needs to be defined, many samples of the spectrum at well-separated frequencies are needed. The spectral lines from quantum transitions, however, must be measured at their specific natural frequencies with allowance for Doppler shifts.

1.2 PROSPECTS FOR FUTURE SCIENTIFIC USE OF THE RADIO SPECTRUM

With the threat of climate change and related environmental changes over the next several decades, the need for environmental information for critical policy and economic decision making will increase. The ability of passive microwave Earth remote sensing to study water in various phases, at both continuum and spectral line frequencies, means that these instruments will be increasingly used to provide key information. Whether obtained for use in day-to-day weather forecasting operations or for long-term climate studies, passive microwave measurements of Earth represent one of the most important scientific uses of the radio spectrum.

A number of contemporary problems in physics also require radio astronomers' continued ability to observe the cosmos. For instance, studies at radio frequencies provide the only way to investigate the epoch of reionization that occurred when ultraviolet radiation from the first stars ionized intergalactic space, bringing the universe to the state in which it exists today. Radio astronomy also provides the only way to study large numbers of pulsars to determine if Einstein's theory of general relativity is actually correct in the universe's most extreme conditions.

Finally, the use of passive radio techniques to observe the Sun provides the prospect of monitoring our own star for subtle changes in emission characteristics that may lead to geomagnetic disturbances. Such disturbances regularly affect the operation of satellite communications, navigation systems, and terrestrial power grids. Just as Earth environmental information is expected to grow in societal importance in the decades ahead, so is space environmental information.

1.3 THE INTERFERENCE PROBLEM

Because the total radio spectrum is a limited resource, competition exists among its various users. In the United States, spectrum use is regulated by the National Telecommunications and Information Administration (NTIA) for federal government users and by the Federal Communications Commission (FCC) for all others. The regulation of spectrum use includes the assignment of frequency bands, the specification of maximum allowed power levels, and other specific conditions ("footnotes") regarding potential interference with other users. The regulations identify the uses as "services" (see §4.1 in Chapter 4), and this report focuses primarily on the Radio Astronomy Service and the Earth Exploration-Satellite Service.

The U.S. regulatory system, as well as the International Telecommunication Union (ITU) system, was organized prior to 1950, when there were far fewer uses for the radio spectrum than there are now.[2] As new technologies have been developed, the FCC has allocated new bands for them. During the past two decades, however, the pace of development in radio communications has begun to strain the regulatory system, with the biggest problem being a lack of unallocated spectrum available for new technologies.

There are two fundamentally different categories of spectrum users. One category consists of active users—those who transmit radio signals to achieve their ends, which may be voice or data communications, radar surveillance, or even Earth remote sensing using radars or other transmitters. As a group, active users need ever-increasing amounts of spectrum for the ever-increasing uses that are invented; telecommunications companies (in particular) pay large sums of money to obtain the rights to use it. The other category consists of passive users, such as those in radio astronomy and passive remote sensing, who operate in receive-only mode. These users also need increasing amounts of spectrum to obtain the increased sensitivity required for new studies and services.

The uses and desires of these two communities of spectrum users are asymmetric, because the passive services do not transmit any radio signals. Accordingly, active users can interfere with passive users but not vice versa. Since the passive services can operate in any spectral band, they can face radio frequency interference from active services over much of the spectrum. This can include (at times) interference in the bands allocated on an exclusive, primary basis to the passive services. Such interference is discussed in Chapters 2, 3, and 4. See Box 1.1 for more information.

[2]For a description of the U.S. regulatory process and the ITU process, see National Research Council, *Handbook of Frequency Allocations and Spectrum Protection for Scientific Uses*, Washington, D.C.: The National Academies Press, 2007, Ch. 1.

BOX 1.1
Passive Service Measurements and Radio Frequency Interference

Receivers for Earth Exploration-Satellite Service (EESS) and Radio Astronomy Service (RAS) activities are extremely sensitive, as they must respond to very weak natural radiations. Following is a list of characteristics of EESS and RAS measurements that must be taken into account in considering radio frequency interference (RFI).

- Technology improvements are enabling more ambitious and sophisticated Earth remote sensing and radio astronomy experiments. Thus system sensitivity requirements—and hence RFI thresholds—are steadily tightening.
- The spectral requirements of the RAS and EESS continue to increase, and some observations in the bands allocated to the active services are essential.
- Weak radio interference can generate erroneous scientific results even when such interference is essentially undetectable. When such interference is detectable, it only becomes so after a long period of observation time has passed, meaning that the entire observation is ruined.
- Radio astronomy bandwidths are large, up to 1 GHz and more, and integration times are often long, up to 10^5 seconds (about a day), and can extend to months.
- Radio astronomy studies extend out to redshifts of greater than 6 (see Box 3.2 in Chapter 3), so that for the most distant objects, frequencies of the spectral lines are reduced by up to a factor of more than 7. For the important hydrogen (H) line at 21 cm (1.42 GHz), for example, this means that sensitive studies need to be made at essentially all frequencies from 1.42 GHz down to the very high frequency range (30-300 MHz).
- Satellite-based passive Earth remote sensing measurements occur on a continuous basis and over the entire globe. A set of line and window frequencies extending from approximately 1 GHz to higher than 500 GHz is used.
- EESS observations of trace gases such as ozone or compounds of nitrogen usually require the measurement of several spectral lines for every molecular species under study. This means that many specific frequency bands are required, and it is not practical to restrict measurements to the bands assigned to the passive services.

To be more specific, three types of unwanted emissions that cause interference are defined in the FCC regulations: “spurious” emission, “out-of-band” emission, and emissions in adjacent channels. See the glossary in this report for greater detail. Loosely speaking, spurious emissions come from a transmitter emitting at frequencies outside its assigned band and are caused, for example, by nonlinearities that generate harmonics. Out-of-band emissions are emissions at neighboring frequencies that are spread into adjacent bands by the modulation process. Both can interfere with radio astronomy and Earth remote sensing observations. The generation of small spurious and out-of-band signals is virtually inevitable due to the technological limitations in transmitter electronics, but the actual levels emitted can be controlled at the transmitter and kept to within allowable limits. Emission in adjacent channels can create a "blocking" interference within a receiver. However, this particular occurrence is a result of the technical limitations of the receiver, not the transmitter. The allowed emission levels within these categories are defined by regulatory agencies (the FCC and the NTIA) and through international treaties, as discussed in Chapter 4. In the vast majority of cases, both spurious and out-of-band RFI is inadvertent, that is, unintentional. Nonetheless, such emissions are prohibited if they rise above the allowed level in a protected band. However, in cases where Earth remote sensing or radio astronomy observations must be made in bands where no primary allocation for these uses exists, there is no recourse with respect to the problems and data outages caused by RFI.

Finding: Owing to their receive-only nature, the passive Earth Exploration-Satellite Service and Radio Astronomy Service, operating from 10 MHz to 3 THz, are incapable of interfering with other services.

1.4 INTERFERENCE MITIGATION

Users of the RAS and EESS go to considerable effort to mitigate the effects of RFI. Such effort includes careful attention to the design of the receivers to block authorized transmissions that are nearby (both in terms of geography and in terms of frequency), excision techniques (in time and frequency) to eliminate unwanted signals, and, now in development, advanced processing techniques to recognize RFI and either excise it or subtract it. These "unilateral" techniques are all expensive to do on a regular basis. Furthermore, there are fundamental limitations on their ability to distinguish natural thermal noise (the desired signal) from an efficiently modulated communications or radar signal (the RFI) in which power is uniformly distributed across the allocated band.

Cooperative interference mitigation involves cooperative use of spectrum, with RAS and EESS users coordinating their observations to take advantage of the large amounts of unused spectrum at any time and location. Cooperative mitigation techniques hold great promise, but are untested and would require new spectrum use policies and practices so that these techniques could develop. Both unilateral interface mitigation and cooperative interference mitigation are discussed in Chapter 4.

Another major mitigation cost is incurred up front when an observatory is located in a remote area to escape RFI. Current interest in locating receiver arrays to the far side of the Moon is perhaps the most extreme example of this type of cost. While this strategy can be useful for radio astronomy, Earth remote sensing satellites observe the entire Earth over the course of each day and therefore could not avail themselves of a similar advantage. See Box 1.2 for more information.

BOX 1.2
Important Characteristics of Radio Frequency Interference

Following is a list of important characteristics of radio frequency interference (RFI):

- Licensed transmitters, such as television, taxi radios, and cellular telephones occupy fixed spectral bands. RFI from these sources can in some cases be eliminated by avoiding those frequency bands. However, vigilance in keeping spurious and out-of-band emissions down to acceptable levels is always necessary.
- Strong spurious and out-of-band signals are in fact seen in Radio Astronomy Service (RAS) and Earth Exploration-Satellite Service (EESS) experiments. For example, Figure 3.10 in Chapter 3 shows an example of interference in the band 1610.6-1613.8 MHz, a band allocated to the RAS on a shared primary basis. Figures 2.15 through 2.18 in Chapter 2 show inadvertent RFI to the NASA-Japan Aerospace Exploration Agency Advanced Microwave Scanning Radiometer-Earth (AMSR-E) sensors at 10.6-10.7 GHz.
- Low-power, unlicensed transmitting devices are rapidly proliferating. They range from cordless telephones to local area computer networks to digital cameras to automotive anticollision radars, to name only a few of many examples. Since these are personal devices, the total emission level is generally proportional to the population and level of development in any given area.
- Radio telescopes gain a great deal of protection from RFI by locating in remote areas—for example, in the National Radio Quiet Zone in West Virginia, behind high mountains, or in remote desert areas. However, the RAS cannot hide from RFI caused by airplanes or satellites flying overhead. The locating of observatories far from commonly used flight paths is considered, when possible.
- The EESS, operating mainly from low-Earth-orbit satellites, cannot escape the RFI caused by multitudes of low-power-radiating devices as it passes over populated areas. Over parts of Europe and North America, some EESS data products are now ruined by RFI.
- Many active communications systems, including television, are moving to more efficient use of spectrum, especially in filling up their assigned bands uniformly. This results in less white space where scientists might be able to operate with passive equipment. It also means that the signals more closely resemble the random noise of natural signals and are thus less recognizable as RFI.

1.5 ENABLING SCIENTIFIC USES OF THE RADIO SPECTRUM

The goal of this report is to highlight the importance of the passive uses of the radio spectrum, to identify issues that threaten the ability of the science services to provide benefits to society, and to recommend steps for the mitigation or elimination of these threats while recognizing the importance of the other services. Chapters 2 and 3 discuss the knowledge gained from and benefits to society produced by the EESS and RAS, respectively, as well as current and future spectrum requirements for maintaining progress. Chapter 4 discusses current trends in spectrum use and technology that shape the environment in which the EESS and RAS operate, as well as methods for mitigating the impact of interference. Finally, Chapter 5 provides the committee's recommendations for continuing to enable passive scientific uses of the radio spectrum. The committee's findings are presented throughout the report and in Chapter 5, and an acronym list and glossary of terms are provided in the appendixes, in addition to other useful supplementary material.

2

The Earth Exploration-Satellite Service

In 1960 the first weather satellite dramatically opened humanity's eyes to the beauty and complexity of Earth's atmosphere. Never before had anyone photographed a hurricane's movement or cyclonic shape (see Figure 2.1) or observed the global form of atmospheric waves on a planetary scale. After proving the usefulness of orbiting weather observations, NASA and the National Oceanic and Atmospheric Administration (NOAA) began developing ever more sensitive and innovative space-based instruments that help people understand the natural world around them and their impact on it (see Box 2.1). Modern observation systems offer economically and societally important forecasts extending farther into the future than ever before, but these advances depend on protected radio frequency allocations.

With the development of more advanced instrumentation, it quickly became clear that there were great opportunities to observe at wavelengths other than what is usually called light. In fact, visible light is now only a small part of the story. Most current satellite sensors also observe terrestrial emissions at infrared and/or radio wavelengths. These environmental applications have evolved over the past 50 years by combining radio astronomy and geophysical techniques to form the new scientific field known as microwave remote sensing.

Human eyes evolved to detect visible light because Earth's atmosphere allows solar radiation to pass through an "atmospheric window" at those wavelengths. In the same way, the "eye" of the satellite (the receiver) is designed to view Earth through atmospheric windows at other wavelengths, rather than observing reflected sunlight as human eyes do. Analogous to what infrared goggles (providing heat

FIGURE 2.1 Hurricane Camille as it approaches the Gulf States in 1969, as photographed from the NASA Nimbus III satellite. Image courtesy of NASA/Nimbus III Satellite.

vision) do, most satellite instruments detect the inherent emission of radiation (heat) from the atmosphere and terrestrial surface at wavelengths that reveal details invisible to human eyes. When the atmosphere itself is of interest, opaque wavelengths that do not pass through the atmosphere but are absorbed by it offer more information. Each window and opaque band responds differently to the various properties of the terrestrial surface and atmosphere, allowing those properties to be studied by a simultaneous analysis at multiple frequencies. The accuracy of these studies increases with the number of observed frequencies. The unique ability of passive microwave sensors to "see through" most clouds makes those sensors essential, particularly where clouds are persistent. The sensors are passive in that they do not transmit signals but instead only receive the natural background emission. Scientists can thus extract information from the radio spectrum on environmental properties as varied as atmospheric temperature and humidity, precipitation rate, soil moisture, ocean salinity, and ocean waves (and therefore surface winds and

BOX 2.1
The Origin of Earth Remote Sensing

Earth remote sensing techniques have developed over many years, evolving out of astronomy and accelerating as satellite technology became more robust.

- *Before 1932:* Use of optical astronomy (initial passive spectral observations of stellar and planetary surface and atmospheric temperatures and compositions, demonstrating basic methods).
- *1932:* The first radio astronomy observations by pioneer radio astronomer Karl Jansky, revealing cosmic radiation.
- *1940-1945:* Wartime studies of centimeter- and millimeter-wave atmospheric absorption spectra and passive radiation; the development of sensitive radiometry.
- *1968:* Launch of the first passive microwave radiometer on the Soviet Cosmos-243 satellite—it observed sea surface temperature, land temperature, snow/ice cover, water vapor, and liquid water using four unscanned window channels at 3.5-37 GHz (unfortunately short-lived, operating only for weeks).
- *1972, 1975:* The first long-lived satellites to image window-channel parameters (humidity over ocean, sea ice, ocean roughness and wind, snow cover, precipitation, land temperature, etc.) and atmospheric temperature profiles: Nimbus-E Microwave Spectrometer (NEMS; two window channels and three opaque channels) and Electrical Scanning Microwave Radiometer (ESMR) imaging at 19.36 GHz launched on the NASA Nimbus-5 satellite in 1972, and the Scanning Microwave Radiometer (a wide-swath imaging version of NEMS) and the dual-polarized ESMR imaging at 37 GHz launched on Nimbus-6 in 1975.
- *1978:* The first operational weather satellites to incorporate imaging passive microwave spectrometers for temperature sounding (microwave sounding unit with four opaque-band channels at 50-58 GHz on TIROS-N and NOAA-6 and -7).
- *1987:* The first operational satellites to monitor surface parameters and atmospheric water (Special Sensor Microwave/Imager with seven window channels at four frequencies, 19.35-85.5 GHz, first launched by the Defense Meteorological Satellite Program).
- *Post-1987:* Launch of continually improved research (NASA) and operational (National Oceanic and Atmospheric Administration and Department of Defense) passive microwave instrument types.

ocean internal waves). The full global coverage provided by satellites enables scientists to monitor Earth's environment far more accurately and completely than had been possible using traditional means such as weather stations and balloon sounders. Satellite data have also greatly improved the accuracy of weather forecasts and enabled sensitive, large-scale climate studies revealing, for example, the effects of ozone-modifying trace gases. Figure 2.2 presents a typical image of the abundance of water vapor over the oceans as observed by combining observations in multiple frequencies obtained by the Advanced Microwave Scanning Radiometer-Earth (AMSR-E) imaging passive microwave spectrometer.

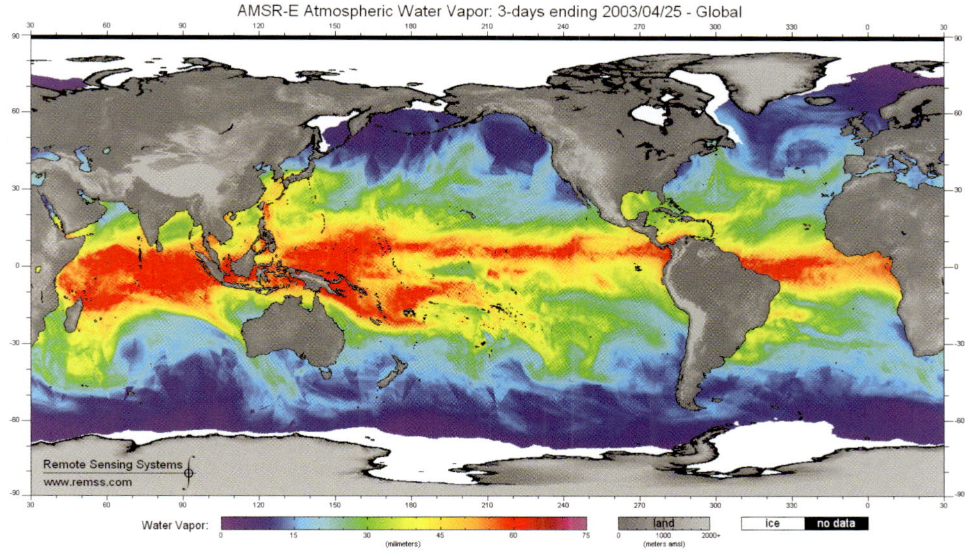

FIGURE 2.2 Advanced Microwave Scanning Radiometer-Earth (AMSR-E) data showing tropospheric water vapor abundance over Earth's oceans, denoted by the colors in the image. Land is denoted by shades of gray, its shade depending on the elevation; sea ice is denoted by white. AMSR-E data are produced by Remote Sensing Systems and sponsored by the NASA Earth Science MEaSUREs DISCOVER Project and the AMSR-E Science Team. Data are available at www.remss.com.

Today, the United States operates a suite of more than 30 satellites that measure Earth's planetary environment and collectively represent many billions of dollars invested by U.S. taxpayers.

The significance of the passive radio services is suggested not only by the substantial government investment in their development and operation, but also by their impact on the national economy. The environmental products facilitated by the passive services are critical for day-to-day, long-term, and severe weather forecasting and also for the Department of Defense (DOD) and for the energy, agriculture, and transportation industries.[1] The U.S. investment in passive Earth observatories provides the nation with a high degree of economic leverage over environmental events.

[1]The 2006 report *Economic Statistics for NOAA* states that "weather and climate sensitive industries, both directly and indirectly, account for about one-third of the Nation's GDP in sectors ranging from finance, insurance, and real estate to services, retail and wholesale trade and manufacturing. Industries directly impacted by weather such as agriculture, construction, energy distribution, and outdoor recreation account for nearly 10 percent of GDP." National Oceanic and Atmospheric Administration, *Economic Statistics for NOAA*, Washington, D.C., 2006.

On a larger scale, Earth's climate is deemed so important to humanity that the 2007 Nobel Peace Prize was awarded to the Intergovernmental Panel on Climate Change (IPCC) and Albert Gore, Jr., "for their efforts to build up and disseminate greater knowledge about man-made climate change, and to lay the foundations for the measures that are needed to counteract such change."[2] The prize was based on the laureates' assessment that large-scale climate change would irrevocably alter living conditions in many places in the world and thus lead to widespread civil unrest. Consistent with this assessment of the importance of climate to humanity are estimates that the potential consequences of global change in its various manifestations (sea ice loss, global warming and drought, coral bleaching, tropical ecosystem collapse, and other interrelated environmental problems) would be associated with unprecedented societal costs to the United States and the world.[3] These staggering costs demand that the most precise information on global environmental processes be made available to decision makers grappling with questions of environmental policy. The precision of this information and the overall understanding of climate change are driven by both observational science and improved understanding and models of the environment, which in turn are dependent on the availability of spectrum for use in environmental observation. At stake are potential measures including limits on emissions of greenhouse gases such as carbon dioxide and methane, limits on aerosols and chlorofluorocarbons, restrictions on deforestation and freshwater usage, and stiff requirements for agricultural and manufacturing practices and the transportation industry.

It is also useful to note the educational value of government programs that apply radio science to environmental problems. These programs are largely conducted either through or in collaboration with universities and thereby train many graduate students at the cutting edge of both radio- and microwave-frequency technology and Earth science, thus contributing to economic sectors critical to U.S. global competitiveness and the defense of the nation.

The importance of environmental radio services has increased in parallel with the use of public and commercial wireless and other electronics technologies (see discussion in Chapter 4). Collectively there has been a substantial increase in the number of human-made radio signals that can interfere with and corrupt needed scientific and operational passive observations of the environment.[4] The commoditization of wireless and other electronics technology has significantly increased the pressure on the passive uses of the spectrum in terms of allocations and disruptive

[2]Available at http://nobelprize.org/nobel_prizes/peace/laureates/2007/; accessed August 26, 2008.

[3]Intergovernmental Panel on Climate Change, Working Group II Report, *Impacts, Adaptation and Vulnerability*, Geneva, Switzerland, 2007.

[4]*Scientific observations* are those conducted for research purposes. *Operational observations* are conducted in consistent, repeated ways for use in products such as weather forecasts.

interference. As quickly as techniques have been developed to mitigate human-made interference, they are eroded by other expanding active uses of the spectrum. Moreover, as the spectral efficiency of wireless technology improves, the interference that it produces increasingly resembles random noise, which is more difficult to identify and mitigate. These difficulties are compounded by the increased use of spectrum licenses that permit unlimited numbers of approved devices to be used, with decreasing means for enforcement or further mitigation. Section 2.5 discusses these difficulties in a variety of circumstances.

Most active services can use coding techniques, better antenna systems, and higher-power transmitters to survive even high levels of interference, but these techniques are not applicable to passive services. There is a fundamental asymmetry between the spectral requirements of active communications services and passive environmental uses. Advances in wireless technology are rapidly increasing the abilities of competing communications services to share radio spectrum through agile time-frequency multiplexing, whereas the measurement precisions of the passive services are intrinsically limited by the strength of natural emissions, the reception bandwidth, and the observing time.

The competition for radio spectrum also has global implications, as the U.S. environment is affected by environmental conditions in other nations and vice versa. Both U.S. and foreign environmental satellites fly over almost the entire globe and continuously observe within the same spectral bands everywhere; thus critical environmental radio bands need to be uncontaminated everywhere. The data from these diverse, Earth-orbiting, multinational assets are increasingly being shared in the global public interest, which parallels the separate national interests, and can be obtained by no other means. Furthermore, the national character of environmental services and the multidecadal times required for their development and use in space make them much less nimble than the private sector that can develop new radiating products in a period of months. It has therefore become clear that a new look at spectrum policies and regulations is necessary in order to protect the critical passive environmental observations by Earth observation satellites and to permit the passive and active services to coexist productively. This chapter discusses the reasons behind the need for new regulations, which are further elaborated on in Chapter 4.

2.1 SPECIFIC APPLICATION AREAS OF PASSIVE MICROWAVE REMOTE SENSING

Earth remote sensing is critically important to the United States and to the advancement of human scientific knowledge about Earth and the environmental processes that support life and commerce. Microwave remote sensing, or the Earth Exploration-Satellite Service (EESS) in regulatory parlance, provides direct economic benefits to the nation by obtaining information that has economic value to both the public and private sectors. In addition, the collection of these data is a highly technical enterprise that strengthens the U.S. industrial, defense, telecommunications, and environmental sectors. The United States operates in a competitive, information-dominated economy that is dependent not only on having access to the passive spectrum for obtaining data for commercial, governmental, and public purposes, but also on having skilled engineers who are trained in the most sophisticated microwave engineering techniques.

Passive and active remote sensing act in tandem to collect environmental information and ultimately to provide the benefits to society referred to above. Much of the data that lead to these benefits, however, is only available from passive microwave sensors, and these sensors have unique needs that must be met to enable the measurements that they make. For example, passive microwave remote sensing is indispensable for better numerical weather forecasting, large-scale monitoring of subsurface soil moisture, and so on, and improvements in weather forecasting are important economically and strategically.

This section presents a sampling of applications in which passive access to the microwave spectrum is essential for the country. The discussion is organized in the following broad topics: weather forecasting and monitoring, severe weather and disasters, climate and global change, resource management, aviation, defense and pubic safety, international partnerships, and education and technology. The subsection on international partnerships includes discussion of a recent international effort, initiated by the United States and the Group of Eight (G8), to ensure that the nations of the world engaged in space remote sensing collaborate in exchanging data to benefit their societies.

Weather Forecasting and Monitoring

Satelliteborne passive microwave sensors are a critical part of the global weather monitoring system. Passive microwave sensors are particularly critical for measuring temperature, humidity, and precipitation profiles in the cloud-affected troposphere below approximately 10 km, where most economically important weather occurs, and in measuring sea surface winds and temperatures and soil moisture. Part of the reason for this importance is that weather radars measure only the reflectivity of water and ice droplets in the atmosphere but are insensitive to these other parameters. Even so, the extraction of useful information from radar reflectivity measurements relies greatly on knowledge of the droplets' size distribution, which requires complex and costly multiband radar measurements to measure directly. Passive microwave radiometers, however, directly measure the total quantity of liquid water as well as water vapor and other variables. Such radiometers can herald impending weather events by measuring the presence of water vapor in advance of cloud formation and then detect the formation of liquid water droplets well in advance of the detection by rain radars. Moreover, when used in conjunction with weather radars, passive radiometers provide a high degree of precision in the measurement of the path- or area-averaged quantities being observed that serve to calibrate the radar's signal. In this manner the radiometer is able to facilitate the radar's capability to provide high resolution. Radars are thus useful in conjunction with radiometers but not as a substitute for them, as exemplified by the recent Tropical Rainfall Measuring Mission (TRMM) and the CloudSat and future Aquarius and Soil Moisture Active Passive (SMAP) missions.

Modern weather forecasts are based primarily on numerical weather prediction (NWP) models run on massively parallel computers. These models use direct data assimilation (DDA), a powerful technique developed during the past two decades that incorporates all available data from satellites, balloons, radars, and surface stations to steer NWP models. Major worldwide centers developing and operating these models are located in the United States, Europe, Canada, China, Japan, and Australia. Their algorithms, from the beginning, have relied heavily on passive microwave measurements of relevant environmental variables, and they will continue to do so as spatial and temporal resolutions improve. Passive microwave data in the opaque temperature-sensitive bands above 50 GHz have been particularly helpful because of their insensitivity to most clouds; these observations probably constitute the single most valuable data source currently enabling 1-week weather forecasts. The demand for improved space and time resolution has been relentless since the inception of NWP modeling in the 1970s and is expected to continue for the foreseeable future, particularly as wireless devices enabled by the Global Positioning System increase the demands for ever more site-specific, personalized information on weather.

In recent decades, the accuracy and usefulness of weather forecasts have increased tremendously because of progress in both NWP systems and satellite-based remote sensing systems. Figure 2.3 illustrates this progress in terms of the number of days for which forecasts of a given quality are obtained. For the highest-quality Southern Hemispheric forecasts, satellite data increase the forecast from 12 hours to 2 days—a factor of four—and for an anomaly correlation of 0.6 the forecast doubles from 3.25 to 6.5 days. The anomaly correlation is a common measure of forecast accuracy, with values above 0.6 generally considered to be significant.

Much of the improvement in forecasting shown in Figure 2.3 is due to the direct use of passive microwave data on their own, and to the integration of microwave and infrared data that combine the best features of both sensor types. Surface wind data over the ocean derived from spaceborne microwave measurements have also been helpful. These improvements are particularly striking in the Southern Hemisphere where data from surface stations and balloon soundings are sparse, but they also extend forecasts in the Northern Hemisphere by roughly

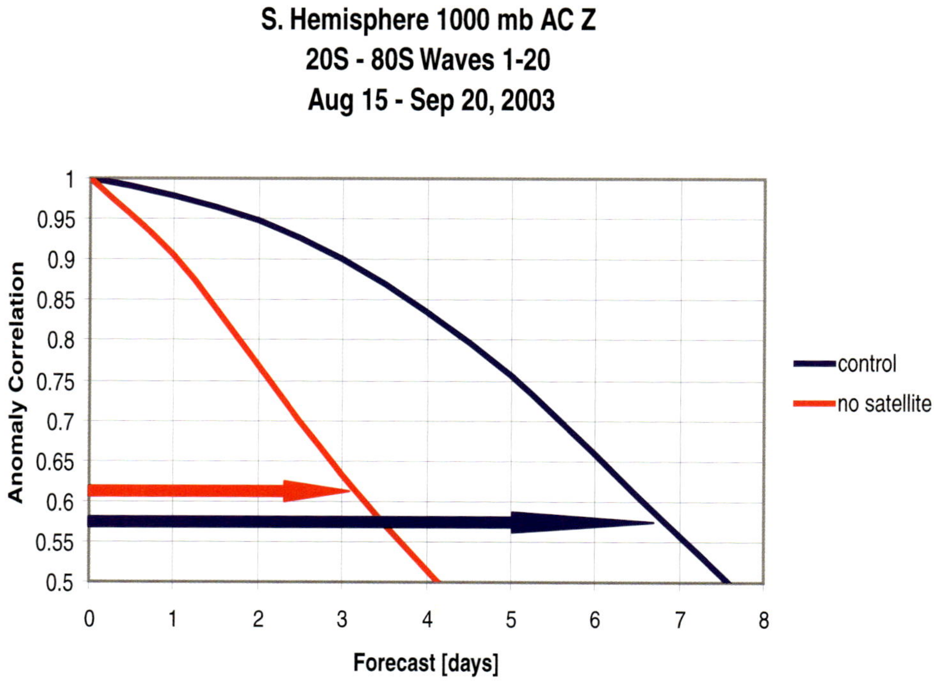

FIGURE 2.3 Anomaly correlation for days 0 to 7 for 500 hectopascal geopotential height in the zonal band 20°-80° South for August/September. The red and blue arrows indicate that the use of satellite data in the forecast model has doubled the length of a useful forecast (i.e., a forecast with anomaly correlation = 0.6). Image courtesy of NOAA.

25 percent. Passive microwave sensors are also useful for tracing the movement of water through normal weather cycles. For instance, surface soil moisture, snow cover, and snow-water-equivalent drive energy exchange with the atmosphere and therefore affect weather forecasts. The major impact of these surface variables on forecast accuracy is just beginning to be seen (Figure 2.4).

Brief discussions of a few specific weather-monitoring topics follow.

Soil Moisture

Accurate knowledge of the parameters of soil moisture (SM) has been shown to improve forecasts of local storms and seasonal climate anomalies. In Figure 2.5, panel (C) shows the observed difference in rainfall between two extreme years, the flood year of 1993 minus the drought year of 1988, over the middle of the United States. Current atmospheric models tend to use sea surface temperatures (SSTs) as their primary boundary condition because so much of Earth's surface is

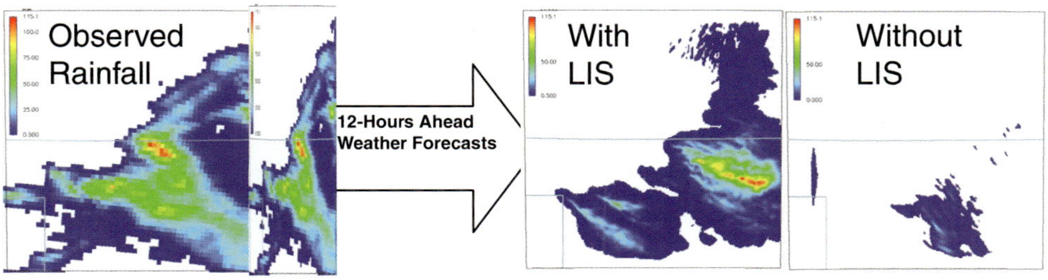

FIGURE 2.4 A depiction of the impact of observations of soil moisture (left) on 12-hour rainfall forecasts that use Weather Research and Forecasting models (for June 12, 2002). Panels at right: Forecasts with and without the Land Information System (LIS) providing improved soil moisture initial and boundary conditions. Image courtesy of NASA.

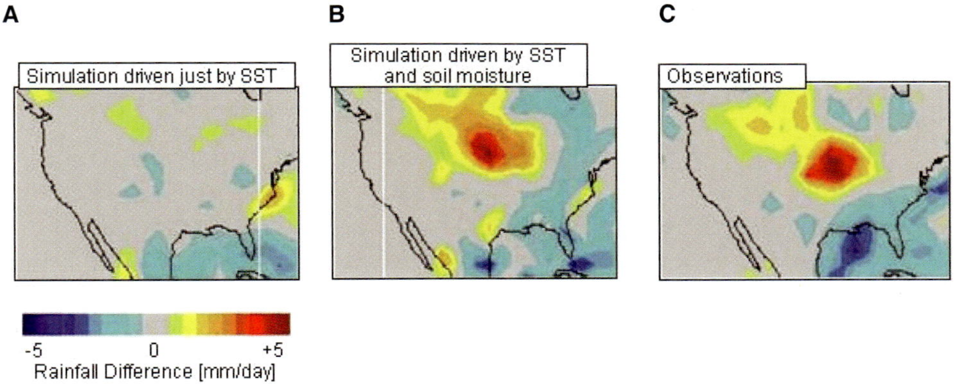

FIGURE 2.5 The value of soil moisture data to climate forecasts. Predictability of seasonal climate is dependent on boundary conditions such as sea surface temperatures (SST) and soil moisture—soil moisture being particularly important over continental interiors. In the results of a simulation driven only by SST (panel A), the climate anomaly in panel C (observed difference in rainfall between the flood year of 1993 minus the drought year of 1988) is not reproduced. Results of the simulation driven by SST and soil moisture (panel B), however, accurately predict this seasonal anomaly. SOURCE: D. Entekhabi, G.R. Asrar, A.K. Betts, K.J. Beven, R.L. Bras, C.J. Duffy, T. Dunne, R.D. Koster, D.P. Lettenmaier, D.B. McLaughlin, and W.J. Shuttleworth, "An Agenda for Land Surface Hydrology Research and a Call for the Second International Hydrological Decade," *Bulletin of the American Meteorological Society,* 80(10): 2043-2058 (1999).

ocean. However, models just using SSTs do not do a good job of capturing seasonal climate anomalies in the middle of large continents. As seen from the results in Figure 2.5(A), the climate anomaly is not reproduced. However, if SM data such as those derivable from space-based 1.4 GHz passive microwave measurements are incorporated, atmospheric models can accurately predict the seasonal anomalies in

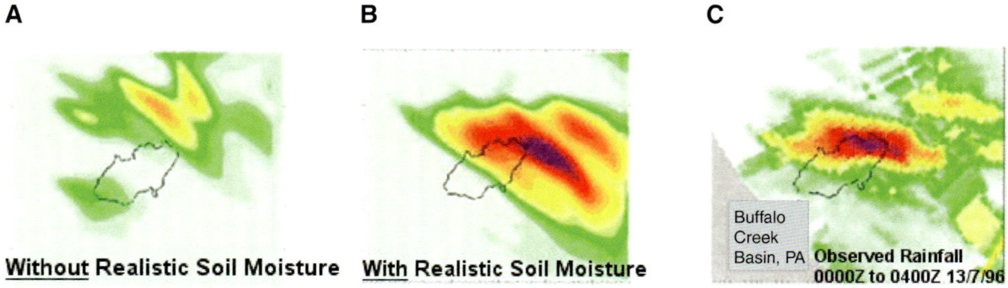

FIGURE 2.6 Soil moisture data improve numerical weather prediction over continents by accurately initializing land surface states. In this example, 24-hour prior forecasts of a high-resolution atmospheric model of rainfall are shown without (A) and with (B) soil moisture input data. The observed data are shown in panel C. Provided by the National Snow and Ice Data Center.

extreme weather, as seen in Figure 2.5(B). In the second example of the importance of SM data (Figure 2.6), NWP can be improved over the continental United States by more accurately initializing the land surface state with soil moisture data.

Soil moisture is also a key parameter in forecasting relating to agriculture, drought, and flooding and for predicting vegetative stress and establishing related government policies. Passive microwave radiometers operating at frequencies of 10 GHz and lower are sensitive to variations in soil density, type, and moisture content and are needed for SM measurements. Radiometry in the 1-2 GHz range is arguably the best means for measuring subsurface soil moisture on a national or global basis.

Sea Surface Winds

Global sea surface wind data are critical for high-quality NWP forecasts, for developing data pertinent to tropical cyclone warnings, aircraft and ship operations, ship routing, and other civil and military operations. Sea surface wind data constitute one of the most important parameters in operational meteorological remote sensing. Space-based remote sensing of sea surface wind vector (SSWV) depends on precision measurements of polarimetric microwave emissions from the ocean surface. These measurements have been shown to improve the forecasting capability of NWP models significantly, thus contributing to maritime and coastal safety and commerce. The accuracy of the wind vector products obtained from the Naval Research Laboratory's (NRL's) WindSat retrievals to date has reached or exceeded the accuracy of the wind vector products available from active scatterometer systems such as QuikScat at moderate to high wind speeds. Also, the ability of microwave radiometers to measure simultaneously atmospheric and sea temperature properties motivates attempts to improve further the accuracy of the radiometer products. In addition, the National Polar-orbiting Operational Environmental Satellite System (NPOESS, which will be the next generation of U.S. weather satellites) will include a microwave radiometer (called the Microwave Imager/Sounder, or MIS) that will likely have many capabilities similar to those of WindSat, including the capability to measure multiple parameters.

Sea Surface Temperature

Global, all-weather sea surface temperature data are critical for NWP and for climate research. SST measurements are important for understanding heat exchange and the coupling between ocean and atmosphere, and SST data are required by operational ocean analyses in order to properly constrain upper-ocean circulation and thermal structure. SST measurements in clear air can be obtained using electro-optical (traditional) instruments; however, clouds prevent these measurements, so passive microwave measurements within the approximately 4-11 GHz region are critical for obtaining coverage in areas that are seasonally cloud-covered. For example, areas in the U.S. Exclusive Economic Zone off the coast of Washington and Oregon are not imaged with traditional satellite SST sensors for weeks at a time owing to persistent stratus cloud cover, necessitating an all-weather solution. The standard SST measurement uncertainty for space-based SST measurements is 0.5 K at 50 km passive microwave (all-weather) capability.

Water Vapor Profiles

Global water vapor profiles are essential to the numerical weather prediction of rainfall and drought and help constrain such predictions in general. As in the case of temperature profile measurements, combined microwave and infrared spectral data can yield what is nearly all-weather global performance, even in most cloudy conditions. Two different types of microwave observations are used: those in transparent bands within which the water vapor absorption stands out (1) against the colder ocean background (ocean partially reflects the extremely cold cosmic background radiation), or (2) against that of cold low-emissivity land. No profile information is usually retrieved—only an estimate of the column-integrated abundance. The frequencies most often used for this purpose include 18.7, 22, 23.8, 31.4, 37, and 89 GHz. To improve retrieval accuracies, these channels are often dual-polarized (horizontal and vertical) and scanned at a near-constant angle of incidence (e.g., TRMM Microwave Imager [TMI], Special Sensor Microwave/Imager [SSM/I], Special Sensor Microwave/Imager Sounder [SSM/IS], WindSat, and AMSR-E). In addition, the opaque water vapor resonance near 183 GHz is often used in combination with some of the lower frequencies; these frequencies generally include 89, 150, 164-168, and 176-191 GHz, but they must be used in combination with temperature profile information to yield the most accurate results (e.g., Advanced Microwave Sounding Unit [AMSU], SSM/IS). Instruments retrieving water vapor profiles are generally used to retrieve other parameters simultaneously, such as cloud water content, precipitation rate, and ice and snow cover information.

Severe Weather and Disasters

One impact of world population growth over the past 50 years is an increased vulnerability to natural disasters. Weather-related disasters include tornadoes, hurricanes, hail, blizzards, floods, mudslides, heat waves, forest fires, and drought. Some disasters have immediate impacts and others long-term effects; for example, rising sea levels could have major impacts on coastal areas, and severe declines in snow cover in the western United States could yield less spring snowmelt and water for summer agriculture and urban needs.

Extreme weather events and other natural disasters can be costly, not only in terms of the immediate loss of life and property that they can cause, but also because of the efforts needed to anticipate and respond to such disasters and in the long-term economic and societal consequences of such events. NOAA estimates that the cost in the United States of damages from tornadoes, hurricanes, and floods alone averages around $11.4 billion annually.[5] Even one major hurricane, however, can significantly exceed these costs. Although the full cost of Hurricane Katrina, which made landfall near New Orleans on August 28, 2005, will not be known for many years, insured losses alone are estimated at $40.0 billion.[6] EESS observations enable significant economic and societal savings owing to their ability to predict such costly natural events and to allow people to prepare for them.

To provide a perspective on the types of costs referred to above, Figure 2.7 highlights major U.S. weather-related disasters over the past 25 years. As growing population density along the coasts of the nation has increased the cost of coastal disasters, the mitigating effect of improved weather forecasting has been reducing those costs by increasing the warning times and the accuracy of the forecasts, leading to increased life-saving evacuation and cost-reducing physical preparations while precluding such steps where they are not needed. According to a report from the National Research Council's Space Studies Board, the error in 3-day-forecast landfall positions of hurricanes was reduced from 210 miles in 1985 to about 110 miles in 2004, arguably halving the preparation area while increasing the population response and preparation effectiveness. Further, the accuracy

[5]National Oceanic and Atmospheric Administration, Office of the Chief Economist, *Economic Statistics for NOAA*, 5th Ed., April 2006, p. 10.

[6]Ibid, p. 18.

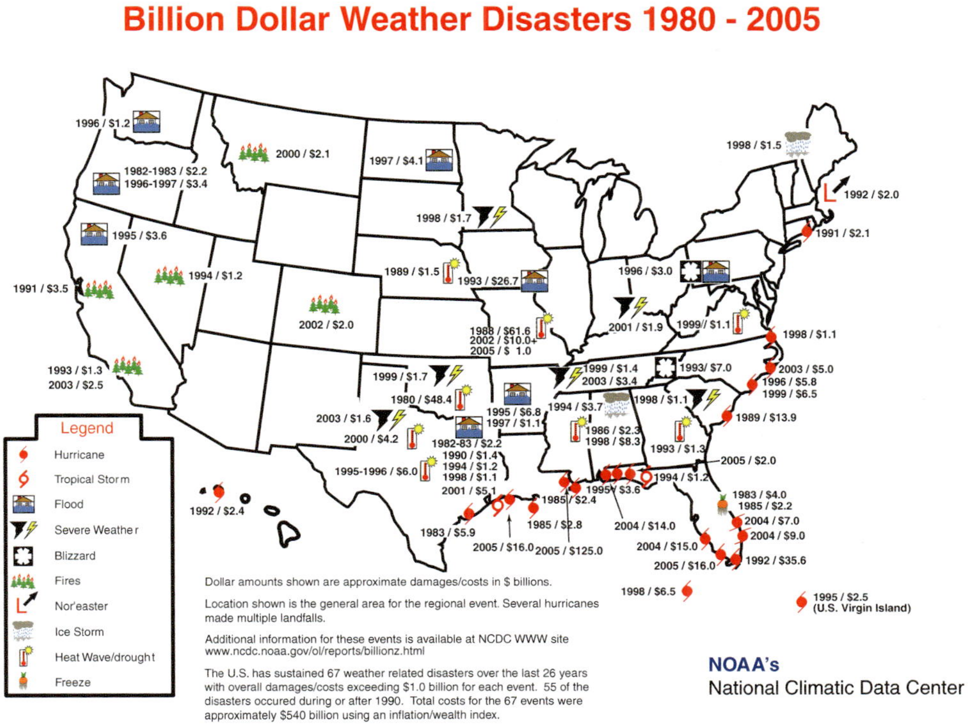

FIGURE 2.7 Billion-dollar weather-related disasters in the United States from 1980 to 2005. NOTE: Earth Exploration-Satellite System measurements are now an important data source for improving the accuracy of forecasts. Image courtesy of NOAA.

of today's 4-day forecasts is about the same as the 2-day forecasts of 20 years ago.[7] EESS measurements have played a major role in improving these forecasts. The insurance industry is also increasingly interested in using passive microwave data to arbitrate claims involving hurricane-related flooding or winds that can often only be distinguished by passive microwave observations.

Figure 2.8 illustrates the value of all-weather microwave SST measurements for hurricane forecasting. Figure 2.8(A) shows how the NASA TMI viewed the cold wake of Hurricane Bonnie through cloud cover as the hurricane moved up the eastern coast of the United States on August 24-26, 1998. Figure 2.8(B) shows the same scene as viewed in infrared by the Advanced Very High Resolution Radiometer (AVHRR) a few days later as Hurricane Danielle moved up the coast on August 27.

[7]National Research Council, *Earth Science and Applications from Space: Urgent Needs and Opportunities to Serve the Nation,* Washington, D.C.: The National Academies Press, 2005, p. 9.

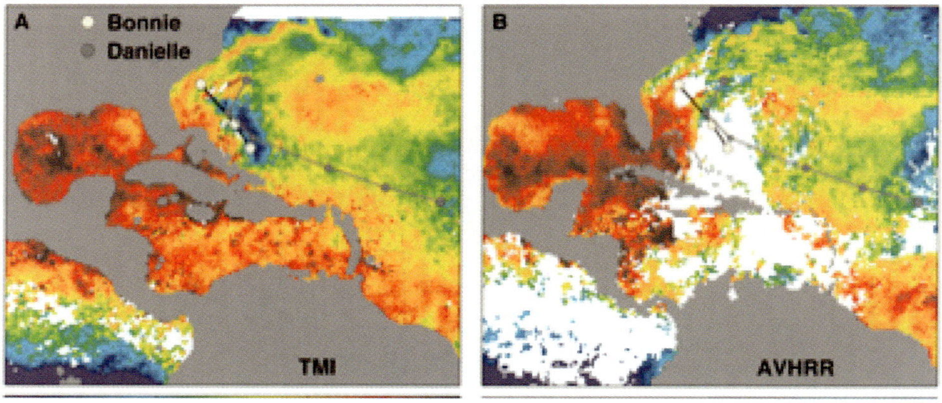

FIGURE 2.8 (A) Microwave imagery at 10 GHz supplied by the NASA Tropical Rainfall Measuring Mission (TRMM) Microwave Imager (TMI) showing a cold wake (blue region near the white dots) produced by Hurricane Bonnie on August 24-26, 1998. (B) The cold wake was not seen by the visible and infrared Advanced Very High Resolution Radiometer (AVHRR) imager because of the areas of persistent rain and cloud cover (white patches) over the 3-day period. When Danielle crossed Bonnie's cold wake on August 29, Danielle's intensity dropped. Although the cloud cover prevented AVHRR from observing this sequence, TMI was able to measure characteristics of the sea surface. Hurricane Bonnie's track is shown by the white dots, and Hurricane Danielle's track is shown by the gray dots. Image courtesy of NASA TRMM Microwave Imager.

The cold wake was invisible to AVHRR because of persistent clouds and rain. A retrospective analysis showed that the magnitude of the cold wake left by Hurricane Bonnie was critical to being able to predict the weakening of the second hurricane, Danielle, a few days later and could not have been done without the microwave measurements of sea surface temperature by TMI.[8] The strong dependence of hurricane growth on local sea surface temperatures makes such measurements through hurricane cloud shields important, particularly since the overturning of the water by the hurricane itself can alter those temperatures rapidly.

An example of the ability of satellite-based passive microwave sensors to observe the high wind speeds of a hurricane is provided in Figure 2.9, an image of the wind speed of Hurricane Katrina as it made landfall near New Orleans on August 28, 2005. In addition, an airborne system, the Stepped Frequency Microwave Radiometer (SFMR), is currently included in NOAA's hurricane-observing research aircraft. Measurements from this system contributed to 23 hurricane advisories in 2005, including the landfall intensity advisories of Hurricanes Katrina

[8]F.J. Wentz, D.S. Gentemann, and D. Chelton, "Satellite Measurements of Sea Surface Temperature Through Clouds," *Science,* 288(5467): 847-850 (May 5, 2000).

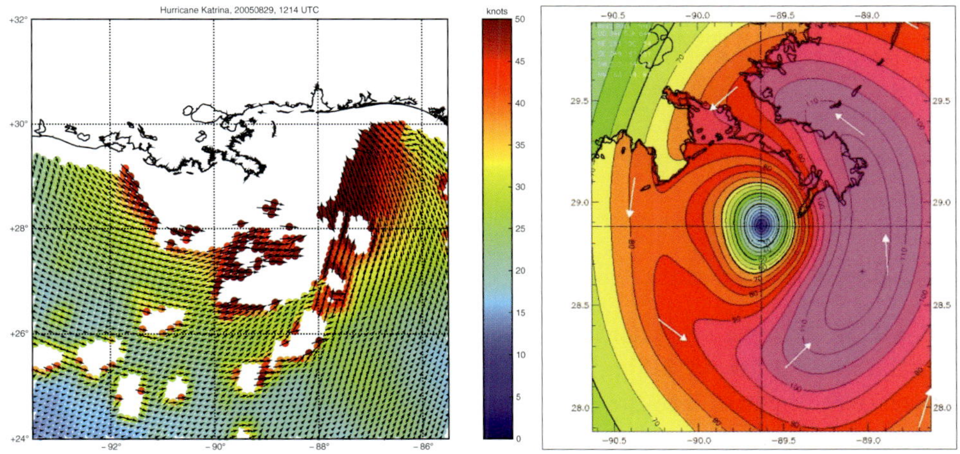

FIGURE 2.9 (Left) Image of the wind speed of Hurricane Katrina (in knots), observed by passive microwave radiometers on WindSat, a Naval Research Laboratory satellite, as Katrina makes landfall near New Orleans on August 28, 2005. (Right) Output from a model that combines data from WindSat and other remote sensing instruments. The model provides information on the hurricane's wind speed. The values over land are extrapolations. Courtesy of the U.S. Naval Research Laboratory.

and Rita. The passive microwave technique is so effective that the U.S. Congress mandated SFMR instruments for the fleet of U.S. Air Force WC-130J Hercules operational weather-monitoring aircraft.

Key impact areas for passive microwave observations of natural disasters include hurricane observations and the forecasting and monitoring of severe regional weather and both drought and flood activity. The general usefulness of passive microwave observations in observing global meteorology also aids in the monitoring of other natural disasters and helps with the associated public-safety requirements.

Climate and Global Change

Perhaps the most significant global issue of the early 21st century is the possibility of global environmental change in response to human activity. The potential consequences of global change in its various manifestations (sea-level rise, sea ice loss, global warming and drought, coral bleaching, tropical ecosystem collapse, and other interrelated environmental problems) can be associated with societal costs from a reduction of 1 to 5.5 percent in global gross domestic product by

2050, depending on the carbon dioxide (CO_2) stabilization level.[9] Such costs demand that the most precise and relevant information on global environmental processes be available to decision makers. In many cases the measurement of key climate-related geophysical variables on a global scale is required, and space-based passive microwave radiometry is often the only reasonable means to collect these measurements.

Atmospheric Temperature Profile and Clouds

Among the most important of human influences on climate is the production of greenhouse gases, including CO_2, methane (CH_4), and various chlorofluorocarbon (CFC; hydrochlorofluorocarbon [HCFC]) and hydrofluorocarbon (HFC) compounds. Of these, CO_2 and CH_4 rapidly become well mixed in the lower atmosphere and affect Earth's radiation budget by trapping infrared radiation that would otherwise be expelled to space. Although CO_2 and CH_4 are themselves potent greenhouse gases and the primary cause of observed global warming, their indirect influence on atmospheric water vapor—a more potent and less predictable greenhouse gas—is perhaps even more important. Tropospheric water vapor provides a feedback mechanism through which increased global warming adds to the capacity of the atmosphere to contain water vapor while simultaneously elevating evaporation rates. The monitoring of water vapor and cloud water content and their effects on global radiation fluxes is thus critical to understanding the causes of climate change and predicting future climates. Currently, cloud coverage and type are the most significant sources of uncertainty in global climate modeling. Radar observations are strongly dependent on unknown drop size distributions, and optical sensors do not penetrate clouds well; thus microwave radiometers on all types of platforms (satellite, aircraft, ships, and ground sites) are essential to making water vapor measurements and thus to the science of climate change.

The ability of passive microwave sensors to observe through clouds, in combination with frequent global microwave measurements of average mid-tropospheric and stratospheric temperatures near 54 GHz, has provided a unique record of global atmospheric change over the past two decades that validates other measures. The observed long-term warming of the mid-troposphere is roughly 0.2 ± 0.04 K per decade.[10]

---

[9]Intergovernmental Panel on Climate Change, *Climate Change 2007: Synthesis Report*, Geneva, Switzerland, 2007.

[10]C.A. Mears and F.J. Wentz, "The Effects of Diurnal Correction on Satellite-Derived Lower Tropospheric Temperature," *Science*, 309(5740): 1548-1551 (September 2, 2005).

Cloud Ice Water Path

Cloud ice water path (IWP) is the vertically summed mass of cloudborne ice particles per unit of area. As ice clouds can reflect a significant amount of sunlight, their impact on global radiative energy fluxes and hence climate change is considerable. Future global IWP measurements from space using passive microwave techniques at frequencies from 89 GHz up to approximately 1 THz could characterize the coupling of the global hydrologic and energy cycles through upper tropospheric cloud processes.[11] Such measurements would enable the development and testing of new self-consistent parameterizations of ice cloud processes and cloud systems, which could in turn guide improvements in ice cloud representation in global Earth system models. These improvements would significantly advance the understanding of the hydrological cycle and climate predictability.

Ozone Depletion and Trace Gases

Climate is also strongly affected by trace gases in the upper troposphere, stratosphere, and mesosphere; some of these trace gases also facilitate the destruction of stratospheric ozone.[12] A diminished ozone layer allows harmful ultraviolet-B (UV-B) radiation from the Sun to reach Earth's surface, where it significantly enhances the probability of the occurrence of basal and squamous cell skin cancers and cataracts. The underlying chemical reactions that cause ozone depletion require chlorine and bromine to be present in sufficient quantities in the stratosphere.[13] This revelation was central to the framing of the 1987 Montreal Protocol on Substances That Deplete the Ozone Layer, which explicitly identified ozone-depleting substances that were subsequently banned in a series of international treaties—in 1989, 1990, 1991, 1992, 1993, 1995, 1997, and 1999. The U.S. Environmental Protection Agency estimated in 1999 that the provisions of the Montreal Protocol, which sought to arrest runaway ozone depletion, would save 6.3 million lives from reduced levels of skin cancer, prevent 299 million cases of nonfatal skin cancers, and prevent the development of 27.5 million cases of cataracts in the United States alone between 1990 and

[11]K.F. Evans, J.R. Wang, P. Racette, G. Heymsfield, and L. Li, "Ice Cloud Retrievals and Analysis with Data from the Compact Scanning Submillimeter Imaging Radiometer and the Cloud Radar System During CRYSTAL-FACE," *Journal of Applied Meteorology*, 44: 839-859 (2005).

[12]J.R. Holton, P.H. Haynes, M.E. McIntyre, A.R. Douglass, R.B. Rood, and L. Pfister, "Stratospheric-Tropospheric Exchange," *Rev. Geophys.*, 33: 403-439 (1995); P.M. de F. Forster and K.P. Shine, "Radiative Forcing and Temperature Trends from Stratospheric Ozone Changes," *Journal of Geophysical Research*, 102: 10841-10857 (1997).

[13]M.J. Molina and F.S. Rowland, "Stratospheric Sink for Chlorofluoromethanes: Chlorine Atom-Catalysed Destruction of Ozone," *Nature*, 249: 810-812 (June 28, 1974).

2165.[14] Passive microwave observations provide a valuable means for monitoring the distribution and concentration of ozone and other trace gases.

Ocean Altimetry and Sea Surface Variables

Microwave remote sensing plays a crucial role in monitoring the global ocean, with radiometry, altimetry, scatterometry, and synthetic aperture radar observations all having important climate applications. Ocean altimetry maps the topography of the ocean surface, from which ocean currents and atmospheric surface pressure can be derived. Maps of the currents are routinely used to help route commercial and military naval vessels and, in the commercial fishing industry, to help locate large fish populations. Sea-level anomalies in the tropical Pacific, derived from altimeters, are perhaps the most sensitive precursor indicators of El Niño and La Niña events up to 1 year in advance.[15] The recent series of satellite altimeter missions—Topography Experiment (TOPEX)/Poseidon, Jason-1, and the Ocean Surface Topography Mission (or Jason-2)—has been able to monitor the rise in global sea level, thus providing an important means of verifying the expansion of the oceans in response to climate change. These missions have also contributed significantly to the ability to forecast the occurrence of El Niño events as much as 1 year in advance.[16] For each radar observation, coincident passive microwave radiometer measurements are needed in order to correct the radar altimeters' determination of sea level for the variations in integrated atmospheric refractivity due to tropospheric water vapor.[17] These refractivity radiometers operate near 19, 23 and 34 GHz and require measurements of brightness temperature that are free of radio frequency interference (RFI).[18]

More generally, the Global Climate Observing System (GCOS) implementation plan[19] includes sustained observations of sea surface temperature, ocean wind

[14]U.S. Environmental Protection Agency, *The Benefits and Costs of the Clean Air Act, 1990 to 2010*, EPA-410-R-99-001, prepared for the U.S. Congress by EPA Office of Air and Radiation/Office of Policy, November 1999, p. 64.

[15]D. Chen, "Applying Satellite Remote Sensing to Predicting 1999-2000 La Nina," *Remote Sensing of Environment*, 77(3): 275 (2001).

[16]D. Chen, "Application of Altimeter Observation to El Niño Prediction," *International Journal of Remote Sensing*, 22(13): 2621-2626 (2001).

[17]S.J. Keihm, M.A. Janssen, and C.S. Ruf, "TOPEX/POSEIDON Microwave Radiometer (TMR): III. Wet Tropospheric Range Correction and Pre-Launch Error Budget," *IEEE Transactions on Geoscience and Remote Sensing*, 33(1): 147-161 (1995).

[18]Ibid.

[19]Global Climate Observing System Secretariat, "Implementation Plan for the Global Observing System for Climate in Support of the UNFCCC (2010 Update)," Draft v1.0, November 13, 2009, Geneva, Switzerland. Available at http://www.wmo.int/pages/prog/gcos/documents/GCOSIP-10_DRAFTv1.0_131109.pdf, accessed 12/30/09.

vector, and total columnar integrated atmospheric water vapor in the list of essential climate variables (ECVs) for satellite-based climate studies. All of these climate variables can be sensed simultaneously through the use of polarimetric, multiple-frequency microwave radiometry, as practiced using NRL's WindSat sensor.

Rainfall and Snowfall Rates

Rainfall and snowfall rates and total amounts of precipitation are highly valuable measurements that can be determined by on-orbit and ground-based microwave and millimeter wave radiometers.[20, 21] Knowledge of these quantities is important to the prediction of floods, of crop health and yield, and of catchment replenishment for hydroelectric, irrigation, and domestic uses, and for other societal benefits and impacts.

Snow

Information about snow and frozen ground is critical for understanding fundamental hydrological processes and for detecting environmental change, assessing its impact, and validating environmental models. Snow cover and snow water equivalent (SWE) data are derived using microwave imagery that is sensitive to emission from different snow depths and structure, in combination with visible imagery. In 2004, a global monthly SWE climatology data set that blended Scanning Multi-Channel Microwave Radiometer (SMMR) and SSM/I passive-microwave-derived SWE with NOAA optical sensor snow maps was completed. The data set serves as an important tool for climate research (see Figure 2.10). Snow cover and SWE are also important parameters for analyzing and improving numerical models of the atmosphere, including surface and atmosphere exchange processes, diagnostics, and forecasting.

---

[20] F. Marzano, P. Ciotti, D. Cimini, and R. Ware, "Modeling and Measurement of Rainfall by Ground-Based Multispectral Microwave Radiometry," *IEEE Transactions on Geoscience and Remote Sensing*, 43: 1000-1011 (May 2005); and F.S. Marzano, D. Cimini, and R. Ware, "Monitoring of Rainfall by Ground-Based Passive Microwave Systems: Models, Measurements, and Applications," *Advances in Geosciences*, 2: 259-265 (2005).

[21] See C. Surussavadee and D.H. Staelin, "Global Satellite Millimeter-Wave Precipitation Retrievals Trained with a Cloud-Resolving Numerical Weather Prediction Model: Part II: Performance Evaluation," *IEEE Transactions on Geoscience and Remote Sensing*, 46(1): 109-118 (2008), which evaluates satellite observations of both rain and snowfall rates from satellites. A good conical scanning reference is C. Kummerow, J. Simpson, O. Thiele, W. Barnes, A.T.C. Chang, E. Stocker, R.F. Adler, A. Hou, R. Kakar, F. Wentz, P. Ashcroft, T. Kozu, Y. Hong, K. Okamoto, T. Iguchi, E. Kuroiwa, E. Im, Z. Haddad, G. Huffman, B. Ferrier, W.S. Olson, E. Zipser, E.A. Smith, T.T. Wilheit, G. North, T. Krishnamurti, and K. Nakamura, "The Status of the Tropical Rainfall Measuring Mission (TRMM) after Two Years in Orbit," *Journal of Applied Meteorology*, 39(12): 1965-1982 (2000).

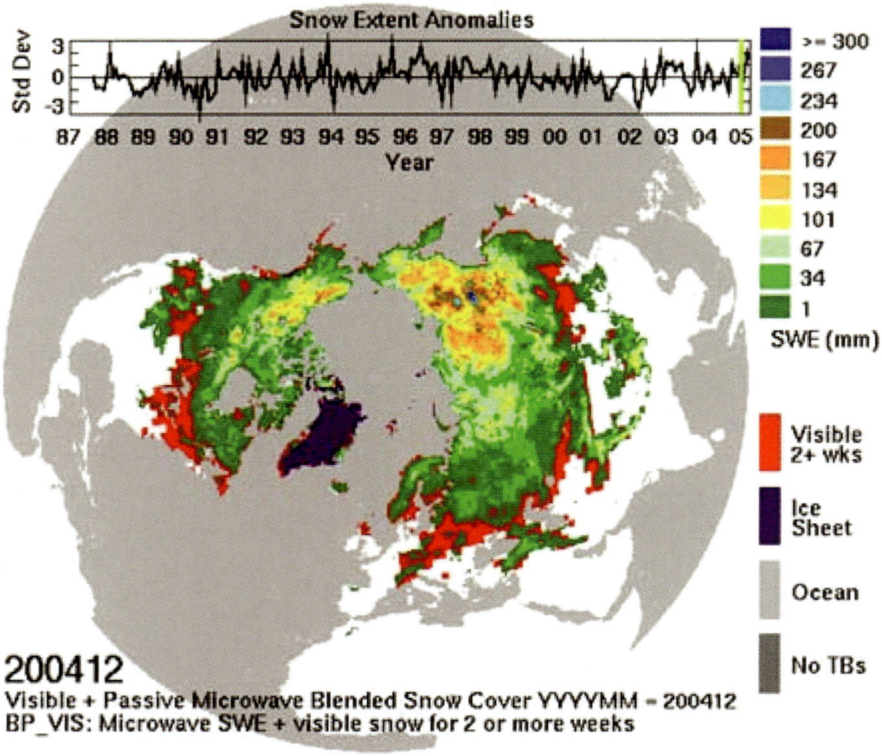

FIGURE 2.10 National Snow and Ice Data Center global monthly Equal-Area Scalable Earth (EASE)-Grid snow water equivalent (SWE) climatology for the Northern Hemisphere, December 2004. The overall data set in which this climatology appears comprises monthly satellite-derived SWE climatologies from November 1978 through June 2003. The global data are gridded to the Northern and Southern 25 kilometer EASE-Grids. Available at http://nsidc.org/research/projects/Armstrong_SWE.html. Provided by the National Snow and Ice Data Center.

Glaciers

Passive microwave sensors can perform spatial mapping of the amount of snow overburden and the melt state of large ice sheets such as those over Greenland and Antarctica. Annual mapping of the ablation zone of the Greenland ice sheet is particularly important as a sensitive means of determining the melt state of the glacial margins and the region of continued deposition of snow.[22] Passive

[22]W. Abdalati and K. Steffen, "Snowmelt on the Greenland Ice Sheet as Derived from Passive Microwave Satellite Data," *Journal of Climate,* 10(2): 165-175 (1997).

microwave window channels from approximately 10 GHz to approximately 90 GHz are sensitive to reflection caused by melting ice water and have been used to study subtle, regionally dependent climate trends in Greenland and Antarctica for nearly two decades. Knowledge of snow overburden is important as a means of estimating the heat transfer from the glacier to the atmosphere, as snow is a good thermal insulator. Passive microwave channels at 18 and 37 GHz are useful for measuring snow depth by virtue of the differential scattering signature available using these two bands.

Sea Surface Salinity

Sea surface salinity (SSS) is a critical missing parameter that scientists need in order to meet climate research goals. Measuring global SSS over time will contribute to scientists' understanding of change in the global Earth system and of how the system responds to natural and human-induced change. Global measurements of SSS can be achieved to approximately 0.2 practical salinity units using space-based passive microwave radiometry at 1.4 GHz and radar scatterometry at 1.26 GHz. These measurements can provide significant new information on how global precipitation, evaporation, and the water cycle are changing. Global SSS variability provides key insight regarding freshwater flow into and out of the ocean associated with precipitation, evaporation, ice melt, and river runoff. Global SSS measurements will also provide important background about how climate variation induces changes in global ocean circulation. The combination of global sea surface salinity and sea surface temperature measurements can be used to determine seawater density, which regulates ocean circulation and the formation of water masses.

Sea Ice

One of the first applications of space-based passive microwave imagery was that of monitoring the location, extent, and thickness of sea ice. The ESMR data set provides the earliest all-weather, all-season imagery of polar sea ice. Some satellite data of sea ice in the visible and infrared wavelengths were available in the late 1960s and early 1970s (before the introduction of space-based passive microwave observations), but because the polar regions are either dark or cloud-covered for much of the year, the generation of consistent, long-term data records from visible and infrared sensing was not practical.

Passive microwave data introduced a major advance in the usefulness of satellite sea ice imaging. The value of passive microwave data for sea ice studies derives from the large contrast in microwave emissivities between sea ice and open water. At 19.35 GHz, open water has an emissivity of approximately 0.44, whereas various sea ice types have emissivities ranging from approximately 0.8 to 0.97. The resulting contrast in microwave brightness temperatures allows accurate estimates of sea ice concentrations (percentages of ocean area covered by sea ice) and hence identification of sea ice distributions throughout the region of observation, as well as temporal variations of these distributions throughout the time period of observation.

Freeze-Thaw Transition

An area of study similar to sea ice involves the seasonal freeze-thaw transition of the Northern Hemisphere, which is a significant source and sink of atmospheric CO_2. The exact timing of the spring thaw and the resulting length of the growing season can fundamentally affect the net carbon exchange budget between land and atmosphere.[23] The thawing of polar tundra also results in more solar absorption and heating, with the possible runaway production of methane from the anaerobic decomposition of subsurface biomass. Passive microwave observations from space—augmented by radar—are the primary means for observing the freeze-thaw transition on a global scale.[24] Determining the freeze-thaw transition requires the use of all of the primary atmospheric window channels between 1.4 and 90 GHz, up to and exceeding the EESS allocated bandwidth on primary or secondary basis.[25]

Biomass

Earth's vegetation canopy, or biomass, is a significant component of the global carbon inventory. It is also a major contributor to the net long-wave/short-wave albedo of the planet and hence to Earth's energy balance and temperature. For these reasons, climate change can both be affected by and can itself affect the global distribution of biomass. The ability to perform comprehensive inventories of biomass from space is recognized as a critical step toward modeling and understanding Earth's climate system.[26] Passive microwave observations operating in all of the

---

[23]S. Frolking, M.L. Goulden, S.C. Wofsy, et al., "Modeling Temporal Variability in the Carbon Balance of a Spruce/Moss Boreal Forest," *Global Change Biology*, 2: 343-366 (1996); J.T. Randerson, C.B. Field, I.Y. Fung, and P.P. Tans, "Increases in Early Season Ecosystem Uptake Explain Recent Changes in the Seasonal Cycle of Atmospheric CO_2 at High Northern Latitudes," *Geophysical Research Letters*, 26(17): 2765-2768 (1999); T.A. Black, W.J. Chen, et al., "Increased Carbon Sequestration by a Boreal Deciduous Forest in Years with Warm Springs," *Geophysical Research Letters*, 27(9): 1271-1274 (2000).

[24]T. Zhang and R.L. Armstrong, "Soil Freeze/Thaw Cycles over Snowfree Land Detected by Passive Microwave Remote Sensing," *Geophysical Research Letters*, 28(5): 763-766 (2001).

[25]National Research Council, *Handbook of Frequency Allocations and Spectrum Protection for Scientific Uses*, Washington, D.C.: The National Academies Press, 2007.

[26]National Research Council, *Earth Science and Applications from Space: National Imperatives for the Next Decade and Beyond* Washington, D.C.: The National Academies Press, 2007.

primary atmospheric window channels between 1.4 and 90 GHz are valuable for monitoring the full range of vegetation canopy water content found in nature and are complementary to optical and synthetic aperture radar techniques. Improved techniques for biomass estimation using passive microwave methods are continuously being developed.[27]

Resource Management

Another application for EESS measurements involves the management of water, energy, and land use, including agriculture and urbanization. All of these applications use observations from multiple sources, including satellites as well as aerial and ground-based measurements. Passive microwave remote sensing is particularly important for assessing phenomena such as soil type and moisture, which can then be related to lake, wetlands, and reservoir storage, to river discharge, and to linkages in the water, energy, and carbon cycles. Other passive microwave products can be used to monitor the size, nutrient status, and other health measures of forests, crops, and vegetation; changes in vegetation type, deforestation, and other land cover; and geographic characterization of the "footprints" of urban areas. Urban and suburban areas play an often-overlooked but important role in Earth's physical and ecological systems, adding to the understanding of mesoscale climatic, hydrologic, and ecologic processes.[28]

Box 2.2 summarizes typical uses of EESS data in reservoir management, the deployment of renewable energy systems, and agricultural forecasting, as reported in a recent evaluation of uses of Earth observations by the U.S. Climate Change Science Program. An additional, long-standing use of data includes the assessment of food security—for instance, in the Famine Early Warning System Network of the U.S. Agency for International Development[29] (see an overview of and details about the network in a report published by the National Research Council in 2007[30]).

One of the most recent applications of EESS data involves the use of information about water quality, vegetation health, population distribution, and other observations as pathways by which to track disease vectors and their implica-

[27] S. Paloscia and P. Pampaloni, "Microwave Vegetation Indexes for Detecting Biomass and Water Conditions of Agricultural Crops," *Remote Sensing of Environment*, 40: 15-26 (1992); G. Macelloni, S. Paloscia, P. Pampaloni, and E. Santi, "Global Scale Monitoring of Soil and Vegetation Using Active and Passive Sensors," *International Journal of Remote Sensing*, 24(12): 2409-2425 (2003).

[28] *People and Pixels: Linking Remote Sensing and Social Science*, Washington, D.C.: National Academy Press, 1998.

[29] Gregory E. Glass, "Rainy with a Chance of Plague: Forecasting Disease Outbreaks from Satellites," *Future Virology*, 2(5): 225-229 (2007).

[30] National Research Council, *Contributions of Land Remote Sensing for Decisions About Food Security and Human Health*, Washington, D.C.: The National Academies Press, 2007.

BOX 2.2
Examples of Earth Exploration-Satellite Service Measurements in Managing Water, Energy, and Agriculture

Reservoir Management

RiverWare is a river basin modeling system that integrates features of reservoirs (recreation, navigation, flood control, and water quality and supply) and electric utility requirements in order to provide basin managers and power managers with a method for planning, forecasting, and scheduling reservoir operations. Inputs to RiverWare include microwave data from the Advanced Microwave Scanning Radiometer-Earth (AMSR-E) and data from other sensors such as the Moderate Resolution Imaging Spectroradiometer (MODIS) and the Advanced Spaceborne Thermal Emission and Reflection Radiometer (ASTER). RiverWare is a collaborative project among the Center for Advanced Decision Support for Water and Environmental Systems at the University of Colorado at Boulder, the Bureau of Reclamation, the Tennessee Valley Authority, and the Army Corps of Engineers.

Renewable Energy Deployment

The U.S. Department of Energy's National Renewable Energy Laboratory uses data from MODIS, the Multiangle Imaging Spectroradiometer (MISR), Advanced Very High Resolution Radiometer, Special Sensor Microwave/Imager, and a host of weather and other data—including measurements of ocean wind, solar and geothermal resources, upper air, and digital terrain/land cover—to assist in the deployment of renewable energy technologies. This model, the Hybrid Optimization Model for Electric Renewables (HOMER), is used to design grid-connected and off-grid renewable energy systems.

Agricultural Forecasting

Agricultural management has long employed moderate-resolution optical imagery, beginning with the Agriculture and Resources Inventory Surveys and continuing with both the Aerospace Remote Sensing and the Large Area Crop Inventory Experiment programs during the 1970s and 1980s. Passive microwave data (from the SSM/I and AMSR-E and other systems) are now routinely incorporated into new agricultural applications. Perhaps most prominent among these is the system run by the U.S. Department of Agriculture's Foreign Agriculture Service (USDA/FAS): the Production Estimate and Crop Assessment Division's Crop Condition Data Retrieval and Evaluation (PECAD/CADRE) system. The FAS collects and analyzes global crop intelligence information and provides estimates to inform official USDA forecasts for the agricultural market. That market includes farmers, agribusiness enterprises, commodity traders, and researchers, as well as federal, state, and local agencies. PECAD/CADRE is one of the largest users of data from EESS agriculture-related measurements.

SOURCE: U.S. Climate Change Science Program, Synthesis and Assessment Product 5.1, "Uses and Limitations of Observations, Data, Forecasts, and Other Projections in Decision Support for Selected Sectors and Regions," November 7, 2008. Available at http://www.climatescience.gov/Library/sap/sap5-1/final-report/.

tions for human health.[31] Glass, in *Future Virology*, discusses the challenges and opportunities provided by EESS data and notes the potential for EESS to advance health assessment beyond the monitoring of disease outbreaks to the forecasting of outbreaks.[32] The advance notice provided by accurate forecasting of outbreaks could allow better deployment of health resources to minimize the spread and impact of disease.

Passive microwave observations of the hydrosphere and the cryosphere are similarly important. A scientific understanding of the mechanism of cycling of freshwater and the amount and distribution of the world's frozen water stores is essential for human survival. Again, passive microwave measurements made at a number of frequencies and from a number of platforms are unique in being able to provide this information.

Aviation

For aviation, the most useful passive services are the U.S. and global weather services and forecasts, which benefit greatly from the inclusion of passive microwave data from satellites. Surface-based, upward-looking microwave radiometers have the unique ability to remotely detect supercooled liquid water that adheres to aircraft flight surfaces and helicopter rotors, and which has been responsible for numerous losses of aircraft and lives. These same radiometers also improve the skills of the forecasters of short-term local aviation weather. Currently, a sparse network of balloon-based profiling (i.e., "radiosonde") sites across the United States, with an average spacing of 315 km, sounds the atmosphere every 12 hours, supplemented by satellite overpasses every several hours. This sparse sampling severely limits short-term forecasting, especially that of severe weather. Ground-based radiometers can duplicate continuously, autonomously, and with minimal ongoing costs many of the data-providing functions of radiosondes (except for measuring winds aloft and providing high vertical resolution).

Fog events have a significant effect on aviation, slowing or halting airport air traffic operations and causing diversions of incoming air traffic.[33] These events are seasonally chronic at some locations and infrequent at others. The onset, duration, and dissipation of fog are difficult to measure and to predict with currently employed technologies (radars, radiosondes, visibility and surface meteorology systems). Ground-based radiometers can measure the vertical profiles of temperature,

---

[31] Ibid.

[32] Gregory E. Glass, "Rainy with a Chance of Plague: Forecasting Disease Outbreaks from Satellites," *Future Virology*, 2(5): 225-229 (2007).

[33] For more information, see "Airline Regulators Grapple with Engine-Shutdown Peril," *Wall Street Journal*, Monday, April 7, 2008.

water vapor, and fog liquid water and therefore have the ability to characterize such fog events. Dubai in the United Arab Emirates has recently had a highly capable three-dimensional fog prediction and monitoring system installed at its airport; the system is based on a microwave radiometer, a wind profiler, surface meteorology, and a computer system.

As an example of the disruptions that fog events can create, on February 15, 2001, a surface-based temperature, water vapor, and cloud liquid water microwave radiometer detected precursors and the onset of meteorological conditions characteristic of persistent ground fog. This fog subsequently shut down Denver International Airport (DIA) for 18 hours at tremendous cost, stranded thousands of passengers, and caused a ripple across the entire air traffic scheduling and flow system. When the situation, including the microwave radiometric temperature, water vapor, and cloud liquid profile data, was replayed into the so-called "MM5" NWP model at the National Center for Atmospheric Research, the model accurately predicted the onset, persistence, and dissipation of this fog.

In another episode, on March 4, 2003, light, freezing drizzle was foreseen, detected, and tracked by a research microwave radiometer monitoring surface-based temperature, water vapor, and cloud liquid water. This condition caused the failure of six jet engines owing to the ingestion of ice on United Airlines (UAL) 737s that were taxiing for takeoff at Denver International Airport, grounding the six aircraft. The direct cost to UAL was reported to be $1.2 million, with an unknown cost resulting from the grounding of the aircraft for engine repairs, other resultant flight cancellations, and further associated costs. In April 2007, this same meteorological condition was foreseen by microwave radiometers, whereupon the radiometer operator unsuccessfully attempted to contact UAL at DIA to forewarn the airline. Two more UAL aircraft lost engines and were grounded. Such losses should diminish as these sensors become operational. To date, UAL has reportedly lost 18 engines at DIA as a result of this meteorological condition. An operational system would have been able to forecast and nowcast this condition, allowing ample warning time for the implementation of preventative procedures.

Beyond aviation issues, fog often causes hazardous surface transportation conditions. It was the cause of a 78-vehicle pileup on Interstate 5 near Fresno, California, in 2002, as well as the precipitator of a number of recent multiple-vehicle accidents across the United States.[34] Ground-based radiometers are being installed in Europe at problem locations for predicting and monitoring fog events.

[34]Gary Sanger, "Winter Weather Summary," available at http://newweb.wrh.noaa.gov/hnx/newslet/spring02/summary.htm; accessed June 22, 2008.

Defense and Public Safety

Although the passive EESS bands are not specifically allocated for defense purposes, they are extensively used by meteorological satellites that support analyses and forecasts serving many defense needs. In fact, many radiometers on operational meteorological satellites are or have been part of the Defense Meteorological Satellite Program (DMSP) satellite constellation operated by the U.S. Department of Defense (as listed in Table 2.2 later in this chapter). For example, the Special Sensor Microwave/Imager Sounder radiometer, which is aboard a DSMP satellite, measures atmospheric temperature and moisture profiles, sea surface winds, cloud liquid content, and land surface parameters on a continuous basis from low Earth orbit. This military meteorological polar-orbiting satellite program has been merged with those of NOAA and NASA in the NOAA Integrated Program Office to form the NPOESS program, which will soon launch its first satellite. Microwave meteorological satellites improve forecasts of the following: (1) weather, which influences essentially all combat missions in the air, on the ground, and at sea; (2) the dispersal and transport of released chemical, biological, or radiological agents, where such knowledge supports defensive measures; (3) the monitoring of the ducting of radio waves over the ocean caused by high gradients in the refractivity of the boundary layer, where such ducting can make ships and aircraft visible to radar at anomalously large distances or invisible at normal distances; (4) the traversability of muddy roads, tundra, or pack ice; (5) battlefield visibility; and (6) trajectory corrections for artillery and other projectiles. In addition there are nonmeteorological covert defense applications of passive sensors: for example, the passive detection of metallic objects, such as tanks and trucks concealed by foliage or camouflage or ships shrouded in fog.

Ground-based microwave radiometers can accurately and precisely measure (to better than 0.5°C in most cases) the temperature profile in the tropospheric boundary layer on a continuous basis. This capability is being used to measure and track inversions that trap clouds, pollution, and aerosols. Knowledge of boundary layer temperature profiles is also important in predicting the transport and spread of accidental or hostile releases of biological agents, nerve agents, and radioactive agents. Such radiometers are being used for continuous monitoring at nuclear power plants in Switzerland, Las Vegas, Beijing, Taiwan, and elsewhere. These radiometers can also measure the water vapor and cloud liquid profile in the boundary layer. Such data are highly important because of the interaction of clouds with aerosols and other gases to form smog. Radiometers can also be used to continuously monitor the atmospheric effects associated with large urban heat islands that can impact health, public utility loads, and human activities.[35]

---

[35]Mikhail N. Khaikin, Iren Kuznetsova, Evgeny N. Kadygrov, and Evgeny A. Miller, "Investigation of Temporal-Spatial Parameters of an Urban Heat Island on the Basis of Passive Microwave Remote Sensing," *Theoretical and Applied Climatology*, 84(1-3): 161-169 (2006).

International Partnerships

It has long been known that sound management of the Earth system, in both its natural and human aspects, requires information that is timely, of known quality, long term in its availability, and global. In 2003, the United States hosted a ministerial-level Earth Observation Summit in Washington, D.C., to promote joint multilateral action that would lead to the continuous monitoring of the state of Earth in order "to increase understanding of dynamic Earth processes, to enhance prediction of the Earth system, and to further implement our international environmental treaty obligations."[36] An ensuing series of summits established a mandate for the development of the Global Earth Observation System of Systems (GEOSS).[37] GEOSS is a complex system of sensors, communication devices, storage systems, and computational and other devices used to observe Earth and to gather the data needed for a better understanding and enhanced prediction of Earth's processes. GEOSS is a "system of systems" consisting of existing and future Earth observation systems contributing to an international and interoperable data network. The emphasis of GEOSS is on societal benefits in nine key areas:

- *Disasters:* Reducing loss of life and property from natural and human-induced disasters;
- *Health:* Understanding environmental factors affecting human health and well-being;
- *Energy:* Improving the management of energy resources;
- *Climate:* Understanding, assessing, predicting, mitigating, and adapting to climate variability and change;
- *Water:* Improving water resource management through better understanding of the water cycle;
- *Weather:* Improving weather information, forecasting, and warning;
- *Ecosystems:* Improving the management and protection of terrestrial, coastal, and marine resources;
- *Agriculture:* Supporting sustainable agriculture and combating desertification; and
- *Biodiversity:* Understanding, monitoring, and conserving biodiversity.

The United States is a key signatory to the development of GEOSS through the international Group on Earth Observations (GEO).

---

[36]Group on Earth Observations, "Declaration of the Earth Observation Summit," Washington, D.C., July 31, 2003, available at http://earthobservations.org/docs/Declaration-final%207-31-03.pdf; accessed December 30, 2009.

[37]See http://www.earthobservations.org/about_geo.shtml; accessed March 31, 2008.

While GEOSS data originate from a variety of sources, there are many important environmental parameters needed by GEOSS users that can be measured only by passive microwave sensors. These include global ocean salinity; sea ice characteristics; soil moisture; rain, cloud, and related atmospheric hydrometric variables; water vapor and temperature profiles under clouds; and trace gases. Without the protection offered by EESS passive radio allocations, the international community would be denied information vital to achieving the goals of GEOSS designed for the benefit of society.

Finding: Passive remote sensing observations are essential for monitoring Earth's natural systems and are therefore critical to human safety, the day-to-day operations of the government and the private sector, and the policy-making processes governing many sectors of the U.S. economy.

Education and Technology

A large number of engineers working in the U.S. telecommunications and defense electronics industry have learned basic radio science skills through graduate or early-career work on any of a number of DOD, NASA, NOAA, National Science Foundation, or Department of Energy passive microwave sensor programs. Examples include spaceborne, airborne, shipborne, and ground-based sensor programs for environmental observation. Not all students trained in the passive microwave area continue their careers in the field, but the importance ascribed to precise instrument calibration, the detection of low signal levels, and innovative signal and image processing provides unusually strong training for careers in many other economically important technology areas. The same can be said for students trained in radio astronomy (see §3.6). Accordingly, Earth remote sensing contributes indirectly to those economic sectors that are critical to U.S. global competitiveness and defense.

In addition to radio science education, the application of passive microwave radiometry to environmental monitoring provides a key means of training Earth scientists. Whether they work in the U.S. government or in organizations around the globe, the next generation of students entering this discipline will need global experience in environmental stewardship and sustainability. The interconnectedness of regions, states, countries, and continents by environmental ties makes U.S. prosperity ever more contingent on the capabilities of environmental scientists, engineers, and managers outside the nation's borders. To this end, valuable global experience in environmental observation is provided through satellite-based passive microwave studies.

Technological spin-offs from passive microwave Earth remote sensing studies are numerous. They include new techniques for instrument calibration, image processing, and data-assimilation capabilities that extend beyond the fields of weather forecasting, radio detection methods using statistical moments, and radio imaging techniques for aircraft navigation in all-weather conditions and for homeland security needs. Additionally, the technology underlying passive Earth remote sensing has led to new submillimeter-wave imaging capabilities for detecting hidden weapons. This technology is now beginning to make its way into screening operations at airports across the United States. Also, the design of cost-effective, stable, integrated microwave receivers has been furthered as a result of needs for such receivers within the passive remote sensing community. Such receiver technology is now found in active communications and radar sensing devices. Finally, the requirement of extremely high main-beam efficiency antennas in passive remote sensing has engendered the development of antennas with low sidelobes for other commercial and defense applications.

Finding: Passive microwave Earth remote sensing provides a diverse and valuable set of educational opportunities.

Finding: In addition to the intellectual benefits that they provide, passive microwave remote sensing studies provide many technological benefits to American society.

2.2 BRIGHTNESS TEMPERATURES, GEOPHYSICAL MEASUREMENTS, AND MISSIONS

Section 2.1 established the range of applications and the importance of passive microwave radiometry. This section describes the processes by which these sensors operate, providing detailed information on the specific geophysical measurements that result. In addition, a summary of previous and future radiometer missions is presented in order to provide context for the current state of passive microwave sensing.

Fundamentals of Microwave Radiometry for EESS Applications

All matter emits low levels of electromagnetic radiation. This radiated "thermal noise" is determined by the temperature and electromagnetic properties of the emitting medium, including its ability to absorb, emit, and scatter electromagnetic waves. Geophysical properties of the medium can be inferred from microwave radiometer measurements of emitted thermal noise power to the extent that those properties are related to the bulk electromagnetic properties of the medium. The thermal noise power radiated at a given frequency is commonly expressed as a "brightness temperature." This is the physical temperature of an ideal emitter (called a blackbody) that would radiate the same amount of noise power at that frequency. The brightness temperature of a scene (reported in units of kelvin by a microwave radiometer) contains the geophysical information of interest. Although multiple geophysical parameters may affect the brightness temperature—for example, the temperature and moisture level of Earth's surface and the temperature, humidity, and cloud properties of the atmosphere—these parameters can be distinguished when they have distinctive frequency and/or polarimetric signatures, so that simultaneous observations of the brightness temperature at multiple frequencies and polarizations enable simultaneous solutions for the geophysical properties of interest. There exists a long history of innovation in passive microwave EESS observations for solving this multiple-parameter estimation problem. For many applications, it is necessary for simultaneous measurements to be made over several octaves of the microwave spectrum in order to distinguish adequately the contributions to the brightness temperature made by the surface and the atmosphere.

Traditional radiometer receivers simply estimate the thermal noise power (in watts) received within a particular radio band by a noncoherent radio receiver consisting of (typically) an antenna, a low-noise amplifier, a filter that limits the observed portion of the frequency spectrum, and a square-law detector that provides a measurement of power in the channel. The output of the square-law detector is averaged over time and then recorded and processed to yield geophysical information.

It is well known that the uncertainty in measurements of brightness temperature is reduced by using larger bandwidths (insofar as is permitted by spectral allocations) and longer integration times (constrained for spaceborne EESS observations by satellite orbit and coverage requirements). While calibration accuracy and internal noise once commonly dominated overall system uncertainty, continuing instrument improvements now often achieve the fundamental sensitivity determined by the time-bandwidth product, thus reaching the maximum achievable sensitivity of the estimated geophysical parameters.[38] Therefore, radio interference to passive systems must be compared to this fundamental limit. In contrast, modern communications systems have yet to approach this so-called Shannon limit. In other words, further improvement in EESS sensor technology will, in general, have minimal impact on measurement accuracy compared to greater time-bandwidth product usage. This is especially true for the most important measurements currently being carried out on an operational basis in EESS. However, technological improvements in the use of spectrum for communications systems are still possible.

[38]See §3.4, "Sensitivity Requirements," for more information on the signal-to-noise ratio of passive microwave measurements.

Although the physics that determines brightness-temperature signatures can be complex, usually just a few principal effects dominate. The absorption and emission of radio waves propagating through the terrestrial atmosphere are strong functions of frequency owing to resonant absorption by atmospheric gases. Figure 1.2 in Chapter 1 shows the total zenith attenuation of microwaves propagating upward through a clear standard atmosphere from sea level. Gas resonances are apparent near 23, 60, 118, 183, and 325 GHz. The primary atmospheric absorbers below 350 GHz are molecular oxygen (resonances near 60 and 118 GHz) and water vapor (23, 183, and 325 GHz). Above the troposphere, absorption and emission by trace gases become more pronounced—for example, HNO_3 at 182 GHz, N_2O at 201 GHz, ClO at 204 GHz, and O_3 at 206 GHz. Radio frequencies used for the Earth Exploration-Satellite Service are usually designated either as "windows" used for observing the surface or total atmospheric attenuation, or as "opaque" and used for estimating atmospheric profiles of temperature or composition. Since radio astronomy uses these same windows to observe the universe from the ground, there is much spectrum compatibility between the two sciences.

Systems for sensing atmospheric properties can be designed to exploit atmospheric absorption and emission resonances. For example, many radiometers include observations near the semitransparent window frequencies 23 and 37 GHz in order to estimate the integrated columnar water vapor and liquid water content of the atmosphere. It is possible to estimate these two unknown abundances because the lower frequency is near the 22.235 GHz water vapor resonance, whereas at 37 GHz cloud absorption is relatively stronger. Observations at the two bands yield two relations that can be inverted to find the two unknowns, that is, the amounts of water vapor and liquid water. It is furthermore possible to estimate atmospheric temperature and/or molecular abundance versus altitude (i.e., temperature or abundance "profiles") by measuring atmospheric brightness temperature as a function of frequency near a resonance. Frequencies in the more transparent regions farther from any resonance generally see deeper into the atmosphere, whereas frequencies near the more opaque core of a resonance sense only conditions relatively near the sensor. Comparing such measurements permits the temperature profile to be determined if the composition is known, or the composition if the temperature profile is known. By combining measurements of different spectral lines, both temperature and composition can be determined simultaneously.

A few underlying physical principles characterize the capabilities of most passive microwave sensors operating in the "window" channels. First, lower-frequency waves generally penetrate intervening media better and sense deeper beneath the surface. Thus low frequencies such as 1.4 GHz are preferred when sensing subsurface soil moisture beneath vegetative canopies. Second, the influence of surface roughness tends to be largest when the length scales of the roughness are comparable to the electromagnetic wavelength. This fact motivates the use of X-band or higher frequencies in attempting to sense the short sea waves (capillary waves) that are most sensitive to sea surface winds at low wind speeds. Third, the dielectric constant of water is a strong function of frequency, temperature, and the water's phase (i.e., ice, liquid, or vapor). A result is that the frequencies most sensitive to sea surface salinity are below approximately 2 GHz, whereas those most sensitive to sea surface temperature lie nearer to 5-10 GHz. Figures 2.11 and 2.12 illustrate for sea and land scenes, respectively, typical sensitivities of microwave radiometers to various environmental properties versus frequency.

Since multiple geophysical properties typically contribute to the observed brightness at a given frequency, multiple frequencies must be observed simultaneously in order to estimate them separately. Because all window channels exhibit some atmospheric absorption and emission, and even atmospheric resonant fre-

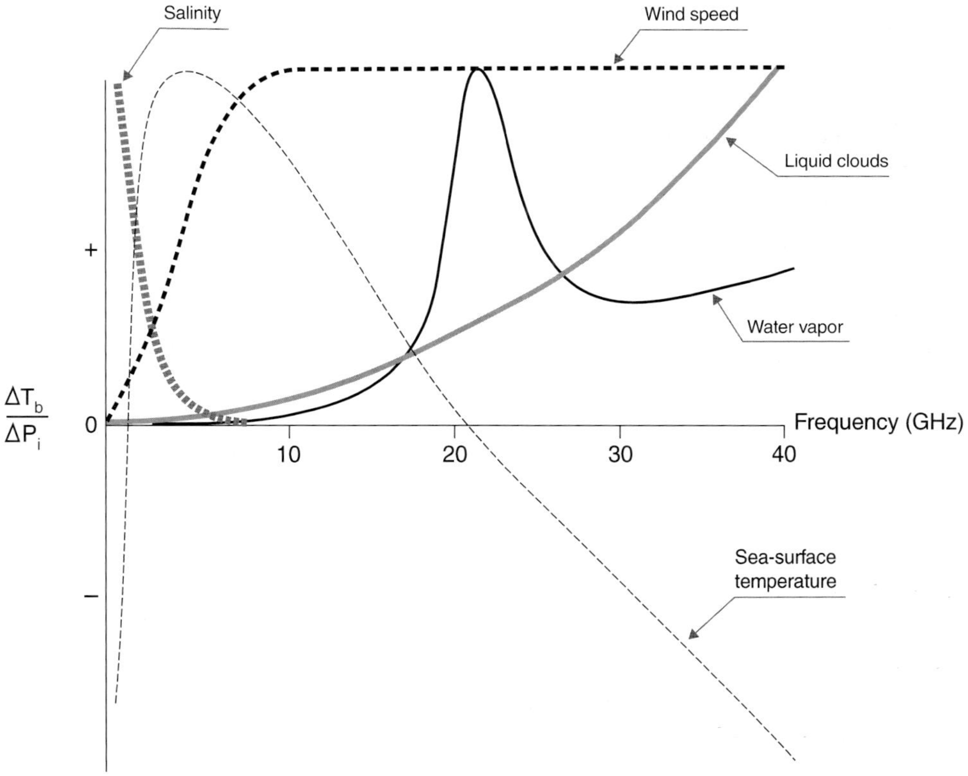

FIGURE 2.11 Ocean scene: relative sensitivity of sea surface salinity, sea surface temperature, cloud liquid water, and integrated water vapor as a function of frequency for space-based measurements. Original figure by Thomas T. Wilheit, NASA-GSFC.

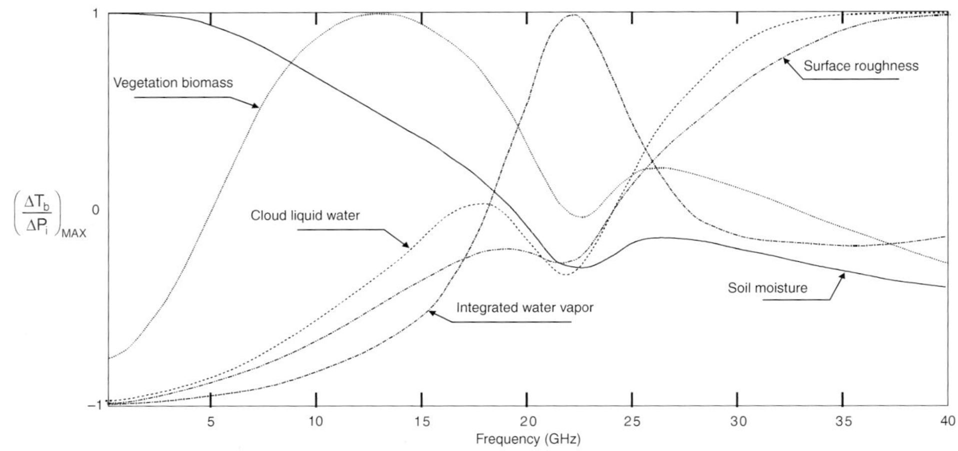

FIGURE 2.12 Land scene: relative sensitivity of the brightness temperature to soil moisture, cloud liquid water, and integrated water vapor as a function of frequency for space-based measurements.

quencies are often not completely opaque, most instruments incorporate both window and opaque channels.

Finding: Effective passive microwave band allocations are necessary for the performance of environmental observation functions.

Finding: Radio wave bands (10 MHz to 3 THz) are indispensable for collecting environmental information associated with specific physical phenomena. Often the same bands are similarly indispensable for radio astronomy, and the passive nature of both services enables them to share the spectrum productively.

Measurement of Specific Geophysical Parameters

Whereas Figure 1.2 in Chapter 1 presents the basic physics of observations through Earth's atmosphere for passive microwave spectral observations, Figures 2.11 and 2.12 also take into account fundamental characteristics of the measured parameters. Because the estimates of the geophysical parameters, also called Environmental Data Records (EDRs), are computed as a function of observed brightness temperatures, it is possible to find the average ratio of a change in a specific EDR to the corresponding change in a particular brightness temperature. This ratio is called the "sensitivity" of the EDR (as distinguished from the radiometric uncertainty of the original radiometer measurement). For example, the sensitivity of surface wind speeds over ocean is expressed in units of $ms^{-1}K^{-1}$. While the numerous channels used in retrieving many EDRs can make this a complicated quantity to determine exactly, the values in Figures 2.11 and 2.12 generally reflect the sensitivity from the primary channels influencing errors in a particular EDR. This ratio permits the accuracy requirements of a particular EDR to be related to the accuracy requirements of the associated radiometric system. Alternately, radio frequency interference levels (K) can be related to resulting errors in EDRs. For example, the sensitivity of sea surface temperature (SST) to the vertically polarized 5 GHz brightness temperature is roughly 0.5 $K(T_b)/K(SST)$. Since current scientific requirements for climate studies include retrievals of SST accurate to within 0.5 K or better, radio frequency interference that causes a 0.25 K change in 5 GHz brightness temperatures would pose a major problem for the retrieval of accurate sea surface temperatures. Similar quantitative statements can be made regarding other EDRs.

It is important to recognize that the EDR products shown in Figures 2.11 and 2.12 are simply unavailable on a global scale from any other type of sensor, particularly for all-weather conditions. These products include critical atmospheric parameters for NWP such as atmospheric temperature, humidity profiles, and precipitation rate. Considering global cloud conditions, surface infrared measurements are possible over an average of 5 percent of Earth's surface and over 30 percent of Earth for the upper troposphere. At somewhat higher altitudes, atmospheric temperature and moisture profiles from microwave measurements (e.g., AMSU) are possible over 70 percent of Earth's surface and 95 percent for the upper troposphere.[39]

Table 2.1 provides a summary of the common geophysical products and the microwave frequencies used for their measurement in current, future, and proposed missions, with an indication of the potential impact of those measurements from radio frequency interference (RFI) based on the current radio frequency environment.

2.3 CURRENT AND FUTURE SPACE MISSIONS, ACTIVITIES, AND SPECTRUM USAGE

Because of the wide range of EESS applications of microwave radiometry, numerous space-based missions are currently in operation or are planned for the near future. Table 2.2 provides a detailed list of such missions, including their planned spectrum usage and intended applications of Environmental Data Records. It is evident that microwave radiometry is widely used by the space agencies of the United States and other nations for sensing both atmospheric and surface properties and that passive microwave radiometry will continue to be widely

[39]R. Saunders, "Use of Microwave Radiances for Weather Forecasting," presentation at the 24th Annual Space Frequency Coordination Group Meeting, France, September 20, 2004.

TABLE 2.1 Summary of Common Geophysical Products and the Microwave Frequencies Used for Their Measurement in Current, Future, and Proposed Missions, Including the Potential Impact of Those Measurements from Radio Frequency Interference Based on the Currently Known Radio Frequency Environment

Environmental Data Record	Role and Significance	Earth Exploration-Satellite Service Passive Microwave Frequencies (GHz):										Summary of Radio Frequency Interference (RFI) Potential
		1.4	6.8	10.7	18/19	23	37	50-60	89	150	183	
Ocean Products												
Sea surface salinity	Global ocean circulation/heat exchange	1.4										High: impact from out-of-band (OOB) emissions and adjacent emissions
Sea surface temperature	Numerical weather prediction (NWP), heat exchange, storm tracking and forecasting, climate operations		6.8									Moderate over ocean: no frequency allocation
Sea surface winds	NWP, storm tracking, operations			10.7 sea	18.7	23.8 (sea)	37					Moderate-low: over ocean, some potential for RFI at 10.7 GHz
Sea ice concentration	Climate, operations				18.7		37					Low: higher frequencies and remote locations
Sea ice age	Climate, operations				18.7		37					Low: higher frequencies and remote locations

Atmospheric Products												
Temperature profile	NWP, storm forecasting, climate							50-60				Moderate-low: potential for RFI due to spectrum-sharing rules at 55-57
Moisture profile	NWP, storm forecasting, climate					23.8		50-60	89	150 166	183	Moderate: collision avoidance radars at 24 GHz
Integrated water	NWP, clouds, climate					23.8						Moderate: collision avoidance radars at 24 GHz
Cloud liquid water	Clouds, storms				18.7	23.8	37					Moderate-low: slightly reduced sensitivity to collision avoidance radars
Cloud ice water	Storms, climate								89	150 166	183	Low: higher frequencies
Precipitation	Operations, climate			10.7	18.7		37	50-56		150 166	183	Moderate-high: 10.7 GHz over land—especially Europe
Land Parameters												
Soil moisture	Climate, NWP forecasting	1.4	6.8	10.7	18.7							High: over land with 1.4 GHz OOB emissions and adjacent emissions and no allocations for 6.8 GHz
Snow water equivalent	Climate, NWP, forecasting				18.7	23.8	37					Moderate-low: some sensitivity to collision avoidance radars (over land)
Surface type	Climate						37		89	150		Low: higher frequencies

NOTE: Additional detail on Earth Exploration-Satellite Service parameters related to this table is provided in Appendix E. In the columns "Earth Exploration-Satellite Service Passive Microwave Frequencies" and "Summary of RFI Potential," red indicates high RFI potential, yellow indicates moderate RFI potential, and green indicates low RFI potential. The colors in between (red-yellow and yellow-green) indicate moderate-high and moderate-low RFI potential.

TABLE 2.2 Past, Current, Future, and Proposed Operational and Scientific Earth Exploration-Satellite Service Missions Providing Critical Operational Data for Weather Forecasting and for Military and Civil Operations in Which the United States Has Participated

Past and Current Missions	Frequency (GHz)	Radio Frequency Interference (RFI) Experiences	Measurements
ESMR (2 past)	19.35	Minimum observed	Sea ice
SMMR (2 past)	6.6, 10.7	6 and 10 GHz, land	Ocean wind speed, integrated water vapor, cloud liquid water, precipitation
NEMS/SCAMS (2 past)	23.8, 31.4, 50-60 (3 channels)		Atmospheric temperature profile
TOVS (MSU) (1 current, 9 past)	50-60	Minimum observed	Atmospheric temperature profile
SSM/I (3 current, 4 past)	19.35, 22.2, 37, 85.5	23 GHz RFI possible from vehicle anticollision radar	Ocean wind speed, integrated water vapor, cloud liquid water, precipitation
SSM/T (3 current, 5 past)	50-60, 89	Minimum	Atmospheric temperature profile
SSM/T2 (4 current, 1 past)	89, 150, 183.31	Minimum	Atmospheric moisture profile
AMSU-A (4 current)	23.8, 31.4, 50-60, 89	23 GHz RFI possible from vehicle anticollision radar	Atmospheric temperature profile
AMSU-B (4 current)	89, 150, 183.31	Significant RFI from nearby spacecraft downlinks	Atmospheric moisture profile
TOPEX (TMR) (1 past)	18, 21, 37	Minimum	Wave mode product correction for ocean altimetry
TMI (1 current)	10.7, 19.35, 23.8, 37, 85	10 GHz, Japan	Precipitation, ocean wind speed, SST*, integrated water vapor, cloud liquid water
JASON-1 JMR (1 current)	18.7 23.8, 34	Minimum	Wave mode product correction for ocean altimetry
HSB (1 current)	89, 150, 183.31	Minimum	Moisture profile
AMSR-E (1 current)	6.9, 10.7, 18.6, 23.8, 36.5, 89	6 and 10 GHz, land	Ocean wind speed, global SST, integrated water vapor, cloud liquid water, precipitation
AMSR (1 past)	6.9, 10.7, 18.6, 23.8, 50.3, 52.8, 36.5, 89	6 and 10 GHz, land	Ocean wind speed, global SST, integrated water vapor, cloud liquid water, precipitation

WindSat (1 current)	6.8, 10.7, 18.8, 22, 37	6 GHz, land; 10 GHz, ocean and land; 18 GHz, ocean	Ocean wind vector, SST, integrated water vapor, cloud liquid water, precipitation
SSM/IS (2 current)	19.35, 22.2, 37, 50-60, 91.6, 150, 183.31	23 GHz RFI possible from vehicle anticollision radar	Ocean wind speed, atmospheric temperature and moisture profile, integrated water vapor, cloud liquid water, precipitation
MHS (2 current)	89, 150, 183.31	Minimum	Atmospheric moisture profile
MLS (1 current)	115.3-122.0, 177.2-206.2, 221.4-240.5, 606.7-667.5, 2481.9-2506.0	Minimum	Atmospheric trace species
Jason-2 AMR (1 current)	18.7, 23.8, 34	Minimum	Water vapor corrections for altimetry
Future Missions	Frequency (GHz)	RFI Susceptibility	Measurements
SMOS (MIRAS) (1)—est. 2009	1.4	High impacts from out-of-band (OOB) emissions	Soil moisture, sea surface salinity
Aquarius (1)—est. 2010	1.4	High impacts from OOB emissions	Sea surface salinity
AMSR-2 (GCOM-W) (1)—est. 2011	6.9, 7.3, 10.7, 18.6, 23.8, 37	6 GHz RFI mitigation	Ocean wind speed, SST, integrated water vapor, cloud liquid water, precipitation
GMI (GPM) (2)—est. 2013	10.7, 23.8, 37, 89, 166, 183.31	10 GHz—European Union	Ocean wind speed, precipitation, integrated water vapor, cloud liquid water, SST*
ATMS (NPOESS) (3)—est. 2013	22.2, 31, 50-60, 89, 166, 183.31	23 GHz RFI possible from vehicle anticollision radar	Atmospheric temperature and moisture profile
SMAP (1)—est. 2013	1.4	High impacts from OOB emissions	Soil moisture and freeze-thaw for weather and water cycle processes
MIS (NPOESS) (3)—est. 2016	6-7, 10.7, 18, 23, 37, 50-60, 89, 166, 183.31	6 GHz, RFI mitigation; 10 GHz—European Union	Ocean wind vector, SST, atmospheric moisture and temperature profile, integrated water vapor, cloud liquid water
SSM/IS (3)—est. 2009	19.35, 22.2, 37, 50-60, 91.6, 150, 183.31	23 GHz RFI possible from vehicle anticollision radar	Ocean wind speed, atmospheric temperature and water profile, precipitation, integrated water vapor, cloud liquid water

continued

TABLE 2.2 Continued

Proposed Missions	Frequency (GHz)	RFI Susceptibility	Measurements
PATH (1)	Microwave array spectrometer	Minimum	High-frequency, all-weather temperature and humidity sounds for weather forecasting and sea surface temperature
SCLP (1)	Ku- and X-band radars; K- and Ka-band radiometer	Possible RFI experience similar to that of WindSat	Snow accumulation for freshwater availability

NOTE: SST (*) indicates reduced capability in colder regions (less than about 12°C). The number of each type of EESS radiometer currently in operation, planned for operation, or proposed for operation is included in parentheses with the listing of its U.S.-based associated missions. Acronyms are defined in Appendix F. In the column "Radio Frequency Interference (RFI) Experiences,"/"RFI Susceptibility," red indicates high RFI potential, yellow indicates moderate RFI potential, and green indicates low RFI potential. The colors in between (red-yellow and yellow-green) indicate moderate-high and moderate-low RFI potential.

employed. Of particular note are the Soil Moisture Ocean Salinity (SMOS) (launch: 2009) and Aquarius (estimated launch: 2010) missions, which will provide the first demonstration of space-based sensing of sea salinity; the Soil Moisture Active Passive mission (estimated launch: 2013 or 2014) for the measurement of global soil moisture; and the NPOESS sensor suite (including the Advanced Technology Microwave Sounder [ATMS] and the Microwave Imager/Sounder system currently being designed) that will provide a wide range of EDR records.

The first U.S. passive microwave radiometer missions date back to 1972. Since then, EESS has continued to fly passive microwave radiometers with ever-increasing capability and covering an expanding range of frequencies. Of note is the current interest in measurements of 1.4 GHz and 6.8 GHz brightness temperature to support sea surface salinity and soil moisture measurements, critical to the continued improvement of weather and climate measurements as described in §2.1, with additional background supplied in Appendix E. In Table 2.2, the number of each type of EESS radiometer currently in operation is included in parentheses with the listing of its U.S.-based associated missions. There is currently a total of 30 missions. Including international missions in which the United States is not involved, the total number is more than 44 missions. Planned and proposed missions include at least 18 more space-based passive radiometers. The complete list of missions represents substantial national and international investment in passive radiometry. Table 2.2 also indicates that several of these new and existing measurements are either currently being impacted by RFI or are highly likely to be impacted by RFI in the near future. A description of the RFI problem at each of the frequencies indicated in either red or yellow in Table 2.2 can be found in §2.5.

Finding: Scientific advances have required increasing measurement precision by passive radio and microwave facilities in order to obtain more accurate and thus more useful data sets. This need for precision will continue to increase.

Finding: Large investments have been made in satellite sensors and sensor networks and in major radio observatories. New facilities costing billions of dollars are under construction or are being designed.

Finding: Radio frequency interference threatens the scientific understanding of key variables in Earth's natural system, now and in the future.

2.4 CURRENT AND FUTURE NON-SPACE-BASED ACTIVITIES AND SPECTRUM USAGE

Although satellites are now the primary data source driving global numerical weather prediction models, over the United States, ground-based meteorological sensors and radiosondes launched at 12 hour intervals from 80 sites have long been the primary source. However, the ever-increasing power of computers leaves NWP models without data between radiosonde sites and launch times, thus limiting models' forecasting capabilities. Moreover, the annual cost per radiosonde launch site is approximately $200,000. To address the problem of cost and sampling density in time and space, less-expensive, continuously operating autonomous ground-based microwave sensors are being developed to augment or replace parts of the present U.S. radiosonde network and thereby to reduce the number of potentially serious unexpected meteorological events that can arise between sample times and places.[40] Moreover, such cloud-penetrating sensors can help calibrate those spaceborne sensors observing water vapor and cloud water content, parameters that vary so rapidly in time and space that they are difficult to validate.

Because of their reliability, economy, and simplicity of deployment, as well as the value of their observations, ground-based radiometers are being implemented in networks in Korea, China, and Europe and are included in a current request for proposals by the National Weather Service. Operational installations are being considered for oil platforms in the Gulf of Mexico. Ground-based radiometers were also deployed around the 2008 Olympics site in Beijing to improve short-term weather forecasts.

Ground-based radiometers can also continuously and locally generate valuable predictive meteorological parameters such as connective available potential energy (CAPE), K-index, total of totals index (TTI), lifted index (LI), and a dozen or so other indices, many of which are associated with severe and sudden-onset weather events.

An example of a current program for the monitoring of global change is the U.S. Department of Energy's Atmospheric Radiation Monitoring (ARM) program, which employs ground-based up-looking passive microwave sensors to characterize the global radiation budget and clouds. These unattended systems continuously measure water vapor profiles and cloud liquid water accurately and inexpensively, relative to radiosondes. Moreover, they provide an integrated measurement that is thought to be more representative of the large-scale behavior of the atmosphere than are the measurements returned by radiosondes. Tropospheric water vapor profiles are measured using a number of bands near the water vapor lines at 22.235 or 183.310 GHz. Bands near the 22.235 GHz water vapor line yield integrated precipitable water vapor (PWV). For 15 years, ARM has used microwave radiometers installed in the tropical western Pacific and at locations up to 70 degrees north latitude for fundamental measurements of atmospheric water vapor. These obser-

[40] Knupp, R. Ware, P. Herzegh, F. Vandenberghe, J. Vivekanandan, and E. Westwater, "Ground-Based Radiometric Profiling During Dynamic Weather Conditions," *Journal of Atmospheric and Oceanic Technology*, 26(6):1057-1073 (2009).

vations have also helped to calibrate radiosondes around the world for weather forecasting and the generation of climate records.

2.5 THE IMPACT OF RADIO FREQUENCY INTERFERENCE ON EARTH EXPLORATION-SATELLITE SERVICE OBSERVATIONS

Microwave radiometers are, by necessity, extremely sensitive radio receivers and are thus very sensitive to radiation from communications, navigation, and other active radio systems. Most radiometers measure total power (brightness temperature) and have no means for distinguishing between naturally emitted thermal noise and the noise-like signals produced by other sources.

Interference can be detected if it is strong enough to be clearly distinguishable from the natural variations in scene brightness temperatures. Lower amounts of interference (i.e., comparable to the geophysical brightness variability) are much more difficult to identify and separate and can therefore compromise the accuracy of the retrieved geophysical information. Although efforts are underway to enhance the abilities of radiometers to detect and suppress interference (as described in Chapter 4), such improvements generally increase costs, data rates, and power consumption while achieving only limited success because of the indistinguishable components of the interference. The following discussion details the process by which human-made sources interfere with radiometry. It presents both specific examples of RFI impacts on Earth observations and justified concerns about future sources of interference.

Introduction to the Problem of Radio Frequency Interference: Immediate Impacts on EESS

EESS radiometers measure the naturally generated background brightness temperature (noise power) of Earth. Since the received power is very small, these radiometers are, by necessity, extremely sensitive instruments. This complicates their design because the background noise temperature that is being measured is so faint that interference power levels of far less than even 10^{-12} W can cause significant measurement errors. Additionally, for spaceborne instruments, the spot size for each individual observation is typically between 12 and 100 km, although smaller spot sizes exist: AMSR-E's spatial resolution is 5 km at 89 GHz. As a result of these spot sizes, pinpointing the precise location of interferers is extremely difficult after launch.

Signals emitted from transmitters operating at frequencies within or adjacent to the passbands of EESS receivers (hereafter to include not only ground-based but also airborne radiometers for Earth observation) are the primary causes of radio frequency interference in EESS measurements. In many cases the interference is due to spurious or out-of-band (OOB) emissions from transmitters operating in bands allocated for other radio services rather than due to signals that are intentionally transmitted in EESS bands. In yet other cases (for example, RFI observed within the 1400-1427 MHz EESS band), it is not always clear whether inadequate filtering within the EESS system or OOB or spurious emissions from active users are the cause, although it is noted that most EESS systems employ state-of-the-art filtering technology that cannot easily be improved.

Spurious and OOB transmitter emissions from commercial devices typically are neither precisely controlled during manufacture nor essential to the devices' intended purposes. The ultimate impact of such emissions on a specific EESS geophysical measurement depends on the sensitivity of the geophysical parameter to changes in brightness temperature, as discussed in §2.2. The high radiometric accuracy and sensitivity achieved by current EESS systems result in commensurately high sensitivity to RFI that can cause errors in the retrieved geophysical parameters. The maximum signal-power contamination that can exist without impacting the information contained in the EESS measurement has been derived by EESS scientists for each of the EESS allocated bands and is documented in International Telecommunication Union-Radio (ITU-R) Recommendation RS.1029-2. Even when false measurements due to RFI are detected and eliminated, forecasts are degraded by the loss of data. Appendix C provides a derivation of the errors in EESS measurements of brightness temperature caused by a collection of anthropogenic sources within the EESS radiometer antenna footprint and frequency passband. Tables 2.1 and 2.2 also provide qualitative assessments of the RFI threat at particular frequencies and for particular missions, respectively.

The RFI threat is especially serious at frequencies lower than 50 GHz, where the atmosphere is largely transparent to radio waves and where frequency bands are widely used by EESS to provide information about environmental parameters. In the first attempts at direct radiance assimilation,[41] only oceanic observations at such transparent frequencies were assimilated into NWP models because the cold microwave background signature of the ocean strongly contrasts with that of the atmosphere. Assimilation of radiances over land at these frequencies was not attempted owing to the relatively poor geophysical signature caused by the high emissivity of land. Recently, though, it has been demonstrated that with increas-

[41] *Direct radiance assimilation* (sometimes just called radiance assimilation) involves the direct use of satellite brightness-temperature measurements to drive the internal state of an environmental model (e.g., a numerical weather prediction model). Now being widely adopted for forecasting purposes, this technique contrasts with the more established technique of performing a retrieval of an environmental parameter using the data. It is generally preferable to retrievals because it uses all available data to achieve the highest forecast accuracy. See, for example, L. Phalippou, "Variational Retrieval of Humidity Profile, Wind Speed, and Cloud Liquid-Water Path with the SSM/I: Potential for Numerical Weather Prediction," *Quarterly Journal of the Royal Meteorological Society,* 122: 327-355 (1996).

ingly accurate land-surface emission models, radiance assimilation at 23.8 and 31 GHz improves both forecasts and quality control of data from other bands. As a result, RFI as weak as 0.1 K or less can limit the use of these bands over land. A similar situation is anticipated with channels in the 1.4 GHz and 6 GHz bands, which are particularly sensitive to surface soil moisture. RFI below 10 GHz threatens to compromise or even eliminate the utility of these bands, which are unique in their ability to provide soil moisture information.

Since ground-based microwave radiometers are valuable for obtaining region-specific temperature and humidity profile data on the lower atmosphere for both nowcasting (typically out to 6 hours) and forecasting, and because they have the unique capability of obtaining low-resolution profiles of cloud liquid water, they are the instruments commonly used in urban areas and at airports where RFI is more likely. However, the tolerable interference levels are quite low for ground-based atmospheric sounding. For example, a 1 W isotropic transmitter at 1 km distance will contribute about 10 K of RFI to a typical uplooking microwave radiometer observing near the assemblage of oxygen lines centered at 60 GHz with a 15 cm antenna aperture, a 300 MHz bandpass filter, and 50 dB antenna sidelobes near the horizon. For a ground-based radiometer, even a 1 K RFI-induced perturbation in a typical seven-channel oxygen band temperature-profiling radiometer can yield an unacceptable 1.4 K error in the retrieved temperature profile. In practice, root mean square (RMS) instrument errors in oxygen-band radiometer measurements are as low as 0.5 K (or lower), and the nominal tolerable RFI level for these systems is 0.05 K. Increasing the number of observation channels in this wave band can mitigate, but not remove, the effect of narrowband RFI.

Evidence of Impact of Radio Frequency Interference on EESS Observations

The corruption of EESS data products by radio frequency interference, including impacts on EESS observations made solely within protected portions of the radio spectrum, has been extensively noted. Typical examples of interference within protected bands and nearby bands follow.

Protected Bands

L-Band (1.400-1.427 GHz)

Observations at 1.4 GHz over land by ground-based and airborne systems in support of remote soil moisture and sea surface salinity estimation are often compromised by what can be identified as OOB emissions from active systems. Total in-band emissions must remain below approximately –140 dBm from 1400-1427 MHz to ensure that anthropogenic (i.e., human-made) emissions do not influence SSS observations to more than a fraction of the necessary stability of 0.05 K that is required to obtain 0.2 psu (practical salinity unit) SSS measurement uncertainty.[42] The RFI contamination that can be tolerated for SM measurement is greater than that for salinity by approximately an order of magnitude; however, the density of transmitters over land is far greater than over ocean. Accordingly, slightly higher RFI contamination levels can be tolerated for 1.4 GHz SM measurements. But in both cases, the maximum tolerable interference level is lower than typical in-band interference from OOB emissions by legal radar transmissions in adjacent spectrum (e.g., at 1.385 GHz). Normal OOB emission limitations determined by the applicable OOB emission mask at 1 percent away from the center bandwidth (e.g., 1400 vs. 1385 MHz) are only slightly below –40 dBc. Using this value, signals within the adjacent EESS band arising from radars within the radiometer antenna footprint can easily exceed the maximum allowed emission level (set at about –140 dBm; see Appendix D).[43]

While few space-based L-band observations have been obtained to date, airborne and ground-based sensors have provided evidence of RFI corruption at levels that prevent geophysical measurements. A recent summary of data measured within the 1400-1427 MHz protected band in April 2005 using the EMIRAD L-band radiometer of the Technical University of Denmark showed significant daily changes in the RFI environment. The percentage of EMIRAD ocean observations impacted by RFI were as low as 1 to 2 percent on most days, but reached 40 to 50 percent in some cases. Repeated occurrences of RFI using the Electronically-Scanned Thinned Array Radiometer (ESTAR) L-band airborne EESS hybrid synthetic-and-real aperture radiometer in the protected 1400-1427 MHz band have been noted in flights over the Eastern Shore region of Virginia in 1999 and over Oklahoma City, Oklahoma, in 1997.[44] These observations have shown clear instances of RFI (Figures 2.13 and 2.14).

A key concern at L-band is the possible influence of long-range air surveillance radar systems in nearby bands. Appendix D presents estimates for the RFI impact on future high-quality soil moisture measurements made by a space-based L-band

---

[42]The stability figure of 0.05 K cited here is a conservative estimate of what is needed to achieve 0.2 psu based on cold water temperatures. D.M. Levine, "Aquarius: An Instrument to Monitor Sea Surface Salinity from Space," *IEEE Transactions on Geoscience and Remote Sensing,* 45(7): 2040-2050 (July 2007), proposes a somewhat higher stability figure of 0.13 K based on measurements averaged over a 7 day window.

[43]N. Skou, S. Misra, S. Sobjaerg, J. Balling, and S. Kristensen, "RFI as Experienced During Preparations for the SMOS Mission," *Proceedings of 2008 URSI General Assembly,* Chicago, Ill., August 9-16, 2008.

[44]D. Le Vine, "ESTAR Experience with RFI at L-Band and Implications for Future Passive Microwave Remote Sensing from Space," *Proceedings of the 2002 International Geoscience and Remote Sensing Symposium. (IGARSS),* Toronto, Ontario, Canada, 2002, pp. 847-849; D. Le Vine and M. Haken, "RFI at L-Band in Synthetic Aperture Radiometers," *Proceedings of the 2003 International Geoscience and Remote Sensing Symposium (IGARSS),* Toulouse, France, 2003, pp. 1742-1744.

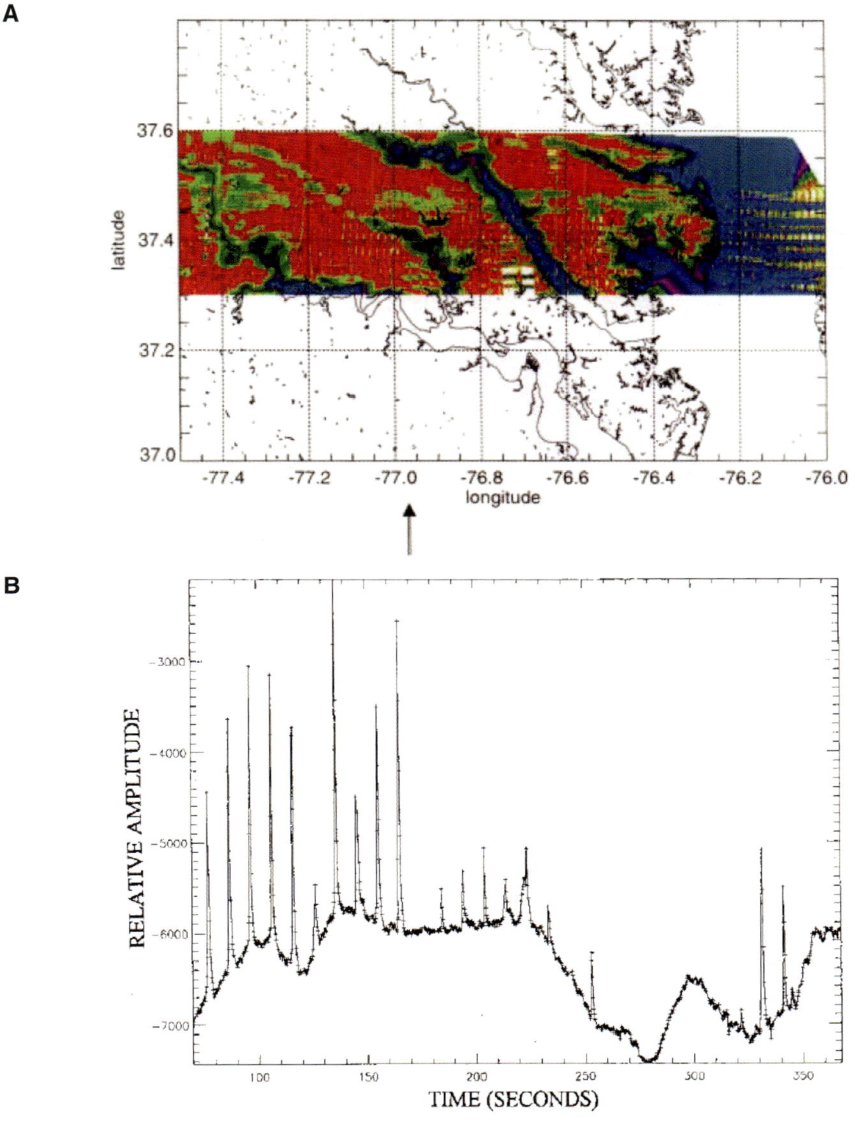

FIGURE 2.13 An example of interference to an airborne EESS radiometer system operating at 1413 MHz from air-traffic radar operating in adjacent segments of spectrum that is possibly due to a combination of spurious emissions from the radar and limitations of adjacent signal rejection in the EESS radiometer. (A) Image from the Electronically-Scanned Thinned Array Radiometer (ESTAR) showing the effects of radio frequency interference (RFI) at 1413 MHz in the vicinity of Richmond, Virginia. The small vertical stripes are artifacts in the image due to strong RFI. (B) The signal is the output of the total power channel. These data were recorded at the location of the arrow in part (A). SOURCE: D. Le Vine, "ESTAR Experience with RFI at L-Band and Implications for Future Passive Microwave Remote Sensing from Space," in *IEEE Int. Geosci. and Remote Sens. Symp. Proc. (IGARSS)*, Toronto, Ontario, Canada, 2002, pp. 847-849, Figures 1 and 2. © 2002 IEEE.

A

B

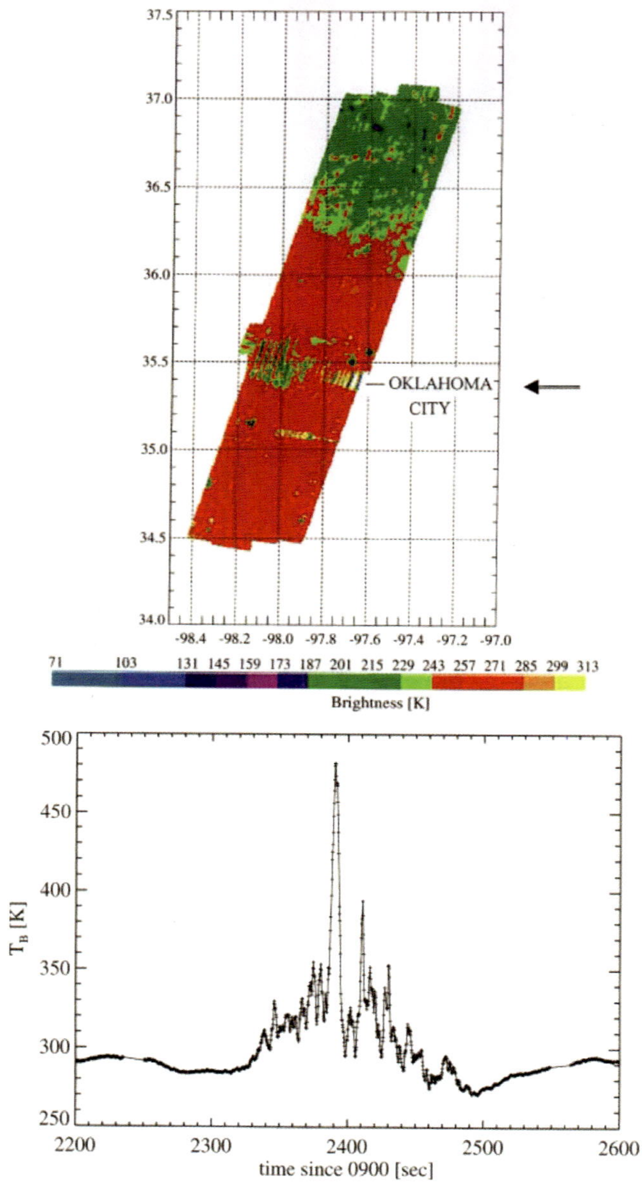

FIGURE 2.14 (A) Electronically-Scanned Thinned Array Radiometer (ESTAR) image at 1413 MHz from the Southern Great Plains experiment (SGP97). The vertical lines west of Oklahoma City are distortions due to radio frequency interference (RFI). (B) Example of RFI in the vicinity of Oklahoma City during SGP97. The signal represents total power and was recorded west of the arrow in part (A). SOURCE: D. Le Vine, "ESTAR Experience with RFI at L-Band and Implications for Future Passive Microwave Remote Sensing from Space," in *IEEE Int. Geosci. and Remote Sens. Symp. Proc. (IGARSS),* Toronto, Ontario, Canada, 2002, pp. 847-849, Figures 3 and 4. © 2002 IEEE.

radiometer, assuming various spurious emission levels at the EESS radiometer at 1413 MHz (the center of this EESS frequency allocation).[45] The results indicate that over the United States where the density of radars is high, RFI would be a significant problem. Synthetic Aperture Interferometric Radiometers (SAIRs) have a wide field of view that, relative to real aperture antennas, increases their vulnerability to strong interference from outside the synthesized antenna beam. Such persistent RFI is a cause for concern for planned space-based EESS systems—for example, the European Space Agency's SMOS sensor.

In both Figure 2.13 and Figure 2.14, it is unclear if the observed RFI was dominated by spurious emissions that fell within the EESS band or by limitations of the EESS passband filtering of emissions in adjacent channels. Regardless, these data demonstrate the need for the mitigation of interference and/or the regulation of OOB emissions radiated in adjacent bands, particularly in L-band. Since the rejection of high-power radar signals in adjacent spectrum is critical to EESS, high-performance front-end filters and other RFI mitigation schemes are essential and have been developed by the EESS community. However, the implementation of filtering schemes, if they are able to suppress RFI to manageable levels, also increases the EESS measurement uncertainty, reduces system sensitivity, increases EESS system cost, and impacts the geophysical data availability. Accordingly, there are practical limitations to minimizing band separation between EESS and active services that need to be considered in developing spectrum usage policy. In addition, in order to design effective RFI mitigation for EESS or prescribe equitable spectrum policy, the interfering signal parameters need to be precisely known. However, only limited information about interfering signals is currently available.

X-Band (10.6-10.7 GHz)

Passive microwave observations at X-band are critical for measurements of sea surface winds (useful for weather prediction and storm tracking) and precipitation (useful for climate and weather monitoring). They are also important for the correction of the effects of land cover on lower frequency (e.g., 1.4 GHz) measurements of soil moisture (useful for climate and weather forecasting). Within X-band, only the sub-band from 10.68-10.70 GHz is protected in the United States and globally for EESS by the ITU, although the wider (and more useful) 10.6-10.7 GHz sub-band has a shared primary allocation within the United States and globally. In addition, observations are also often made including the adjacent sub-band 10.7-10.8 GHz, or including even wider sub-bands on an as-available basis with active services. An example of the use of a wider total band is the Naval Research Laboratory's WindSat sensor, which uses 10.55-10.85 GHz.

[45] Ibid.

Currently, X-band passive microwave imagery over North America appears to be free of obvious RFI from anthropogenic emissions, as illustrated by the example in Figure 2.15 from AMSR-E. The EESS measurements in this band required the use of the full allocated bandwidth of 100 MHz (10.6-10.7 GHz). It is important to note that all but the top 20 MHz of the EESS allocated band is shared with the Fixed Service (FS; point-to-point transmissions, such as radio relay towers); thus, based on Figure 2.15, it appears that U.S. frequency assignments have avoided the 10.6-10.68 GHz segment, which has been beneficial to EESS. However, as the need for spectrum for active services continues to expand, there is concern that significant usage of the 10.6-10.68 GHz band (currently shared with FS) could lead to a scenario at X-band that would resemble the worsening RFI environment at C-band observed between 1987 and 2003 (depicted in Figure 2.23 later in this chapter). A comparable degradation at X-band would be highly detrimental to EESS measurements and their associated data products. Similar concerns also exist at K-band (18.6-18.8 GHz), wherein EESS measurements have begun to display occasional RFI, as can be observed in WindSat imagery (see Figures 2.19 and 2.20 later in this chapter).

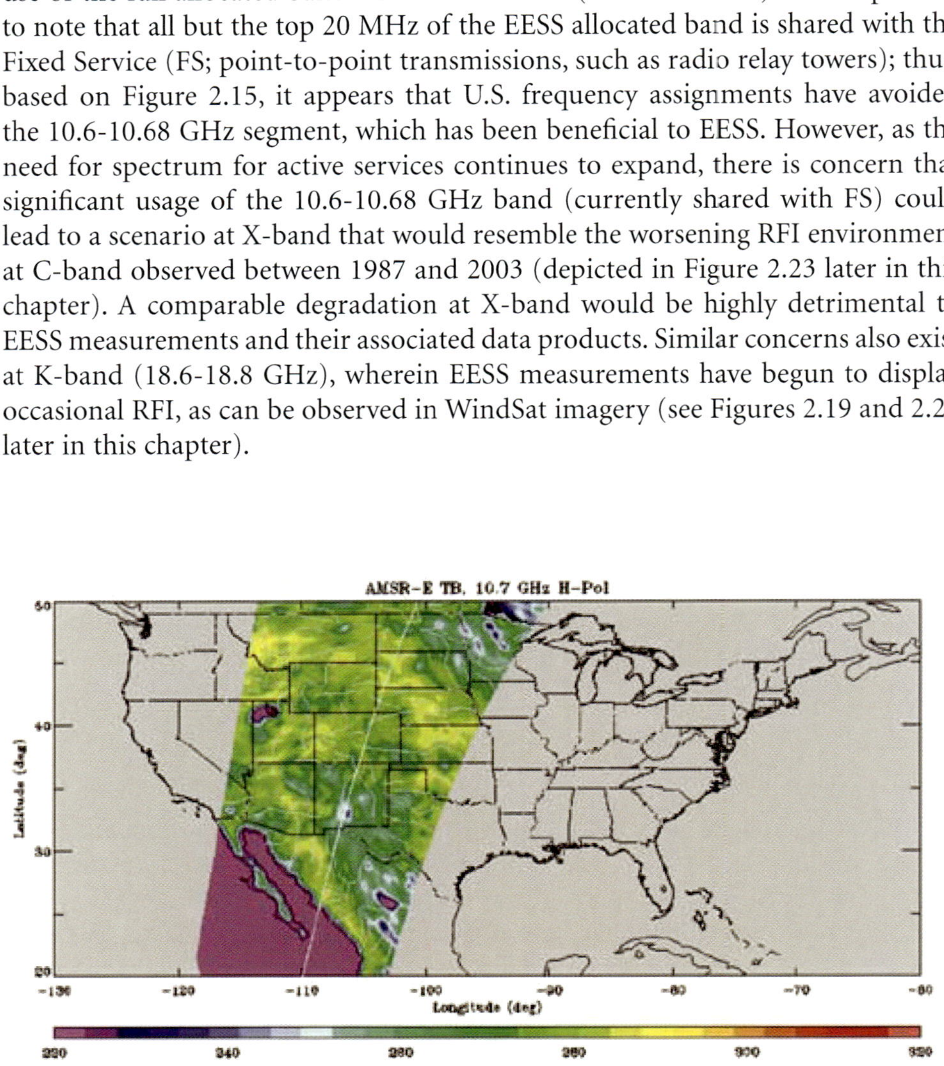

FIGURE 2.15 Brightness temperature as measured by the Advanced Microwave Scanning Radiometer-Earth (AMSR-E) at 10.6 GHz with horizontal polarization over the United States. This observation appears to be free from interference. (L. Li, E. Njoku, E. Im, P. Chang, and K. St. German, "Frequency Interference over the U.S. in Aqua AMSR-E Data," *IEEE Transactions on Geoscience and Remote Sensing,* 42(2): 380-390 (February 2004), from Figure 1.) AMSR-E data are produced by Remote Sensing Systems and sponsored by the NASA Earth Science MEaSUREs DISCOVER Project and the AMSR-E Science Team. Data are available at www.remss.com.

RFI in global 10.7 GHz brightness-temperature measurements was first detected by the TMI radiometer in 1997 during observations over both urban and remote locations of Japan. Subsequently, AMSR-E, launched in May 2002, showed substantial RFI in several European locations that were not observable by TMI due to its near-equatorial orbit (Figures 2.16 and 2.17). Currently, about 2 percent of the land area of Europe is unavailable to AMSR-E for measurements at 10.7 GHz, and an unknown, larger fraction may be adversely affected below the threshold of obvious detectability. However, the looming problem of RFI at X-band is not confined to land areas. Data at 10.7 GHz, such as those provided by WindSat and AMSR-E for SST, ocean wind, and maritime precipitation measurements, often experience substantial RFI from geostationary transmitters operating immediately adjacent to the upper edge of the 10.7 GHz EESS band segment. This maritime RFI is caused by downward-propagating geosynchronous broadcast signals reflecting from the ocean surface into the antenna beam of the EESS sensor. The RFI results in areas of the Mediterranean Sea, the eastern Atlantic Ocean north of the equator, and

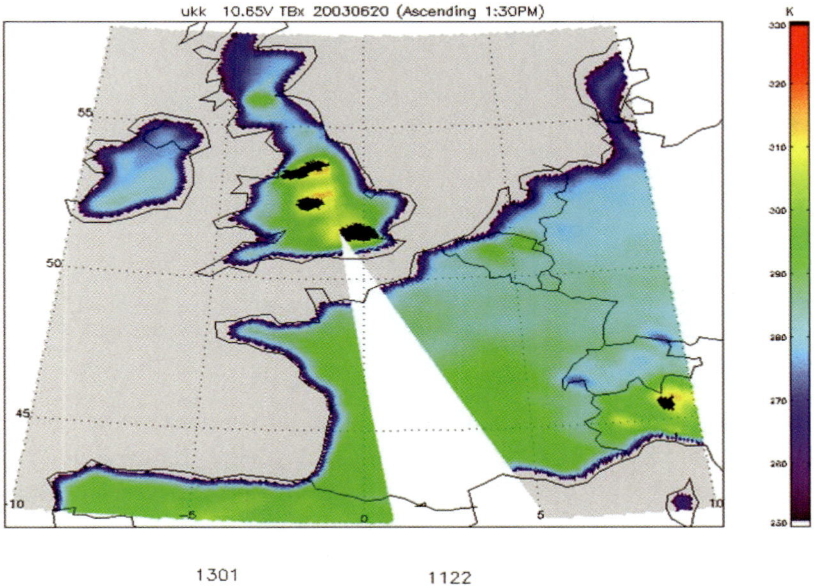

FIGURE 2.16 Passive microwave imagery from the Advanced Microwave Scanning Radiometer-Earth (AMSR-E) on the NASA Earth Observing System Aqua at 10.7 GHz over Europe. Strong emissions over the United Kingdom and portions of Italy are seen as saturated brightness temperatures (black spots). These areas, and nearby yellow and red areas in this example, cannot be used for the retrieval of geophysical parameters such as soil moisture, precipitation, and cloud water. AMSR-E data are produced by Remote Sensing Systems and sponsored by the NASA Earth Science MEaSUREs DISCOVER Project and the AMSR-E Science Team. Data are available at www.remss.com.

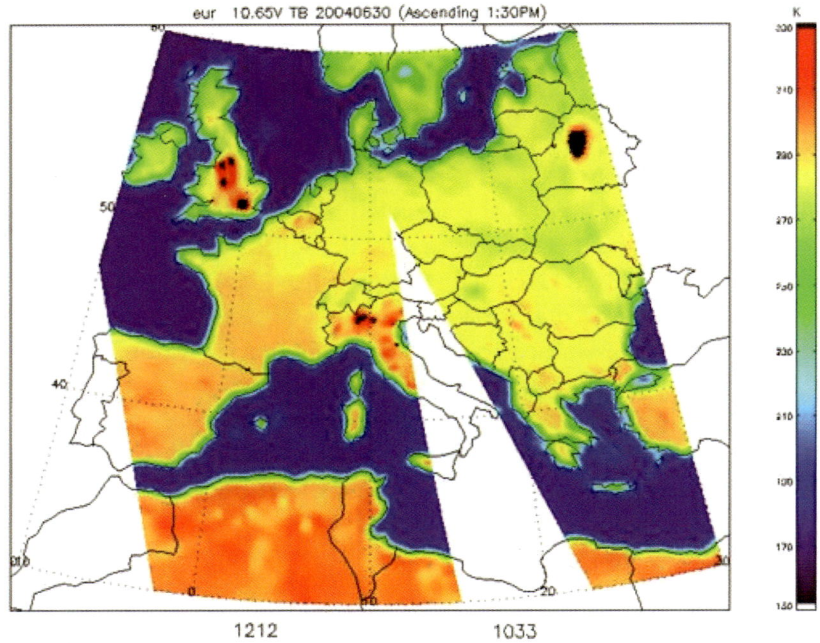

FIGURE 2.17 Expanded region of Europe shown by the Advanced Microwave Scanning Radiometer-Earth (AMSR-E) brightness temperatures at 10.65 GHz indicating the dependence of radio frequency interference (RFI) on political boundaries. RFI can be seen in England, Italy, and Belarus, whereas other countries appear to show none. These instances show the critical role of informed frequency managers and assigners within their respective jurisdictions for limiting impact between services of shared spectrum segments. AMSR-E data are produced by Remote Sensing Systems and sponsored by the NASA Earth Science MEaSUREs DISCOVER Project and the AMSR-E Science Team. Data are available at www.remss.com.

the western Atlantic off the coast of Brazil being unavailable for sea surface wind, temperature, and heavy rain measurements, as shown in Figure 2.18.[46] Southerly views of upwelling microwave brightness temperatures are typically measured by polar-orbiting EESS satellites in the descending phases of their orbits, so such RFI is typically observed in half of all such data over the Mediterranean Sea. The problem also manifests itself as RFI-corrupted calibration views of what should otherwise be cold space during portions of the WindSat orbit.

Analysis of the WindSat polarimetric channels has shown that significant RFI is occurring within the sub-band 10.55-10.85 GHz. Based on earlier measurements

[46]Hotbird 4 channels 110 (10.71918 GHz), 111 (10.72713 GHz), and 112 (10.75754 GHz) are likely candidates.

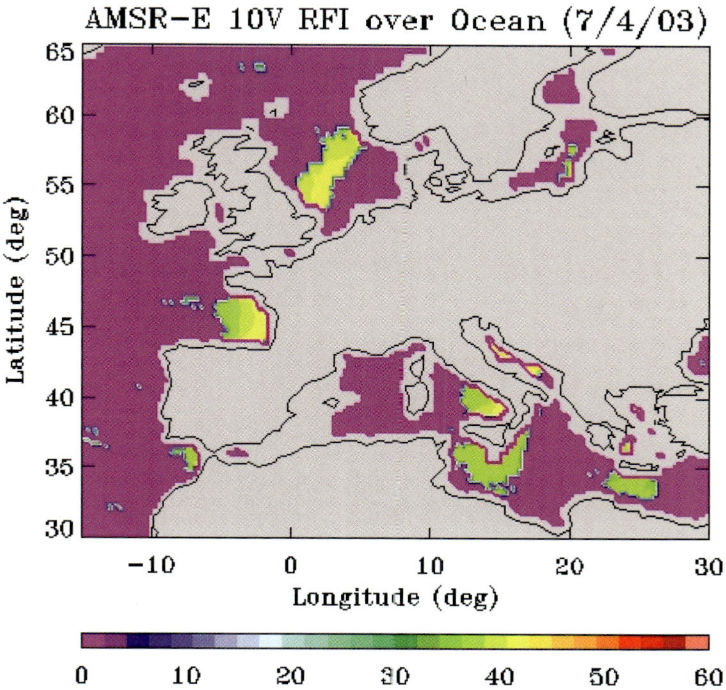

FIGURE 2.18 Example of radio frequency interference (RFI; areas in green and yellow) occurring at X-band from oceanic reflections of geosynchronous broadcasts in bands adjacent to those observed by the Advanced Microwave Scanning Radiometer-Earth (AMSR-E). In this example AMSR-E is operating in the EESS band 10.6-10.7 GHz and is experiencing perturbations higher than 40 K in measured brightness temperature during its descending phase. This level of RFI is far greater than approximately 0.2 K, the minimum level of perturbation that degrades environmental models that use sea surface temperature data derived from AMSR-E. AMSR-E data are produced by Remote Sensing Systems and sponsored by the NASA Earth Science MEaSUREs DISCOVER Project and the AMSR-E Science Team. Data are available at www.remss.com.

using SMMR compared with recent measurements using WindSat, strong X-band RFI in Europe and Japan appears to be increasing over time. The X-band channels of the airborne Polarimetric Scanning Radiometer (PSR) have also detected RFI over the United States, although to a lesser degree than at C-band. The high-resolution PSR mapping capabilities permit pinpointing the location of sources of RFI, but only within limited data sets.

Finding: Whereas most frequency regulations for active services are defined on local or regional bases, passive EESS observations are global by nature. As a result, a high level of international cooperation is required to maintain and enforce passive allocations.

K-Band (18.6-18.8 GHz)

The 18.6-18.8 GHz band is a critical resource for EESS that supports many operational environmental products, such as snow cover, sea surface wind speed, and soil moisture measurements. Snow water equivalent measurements, which are increasingly important for water management, specifically require the use of observations at a frequency near this band. Thus there is a global primary allocation for EESS at 18 GHz.

Evidence of RFI has been found in 18 GHz WindSat space-based observations, as shown in Figure 2.19 for the Paris and London metropolitan areas. Sparse but recurring RFI at 18 GHz has been observed on nearly every continent, as shown in Figure 2.20. As a result, scientists are concerned that increasing use of the spectrum near 18 GHz will increase RFI for WindSat and other EESS radiometers.

K-Band (23.6-24.0 GHz)

Space- and airborne radiometric observations of the weak water vapor resonance near 22.235 GHz are at risk owing to recent rule changes that allow automo-

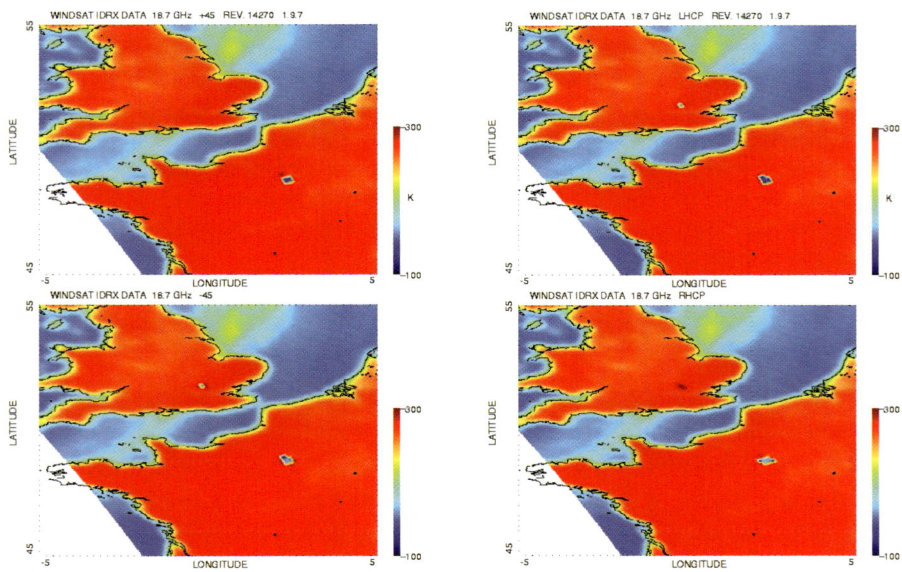

FIGURE 2.19 Brightness-temperature data from the WindSat 18.7 ±45° channels (both left) and 18.7 R/LCP channels (both right) showing strong radio frequency interference over Paris and London. Courtesy of the U.S. Naval Research Laboratory.

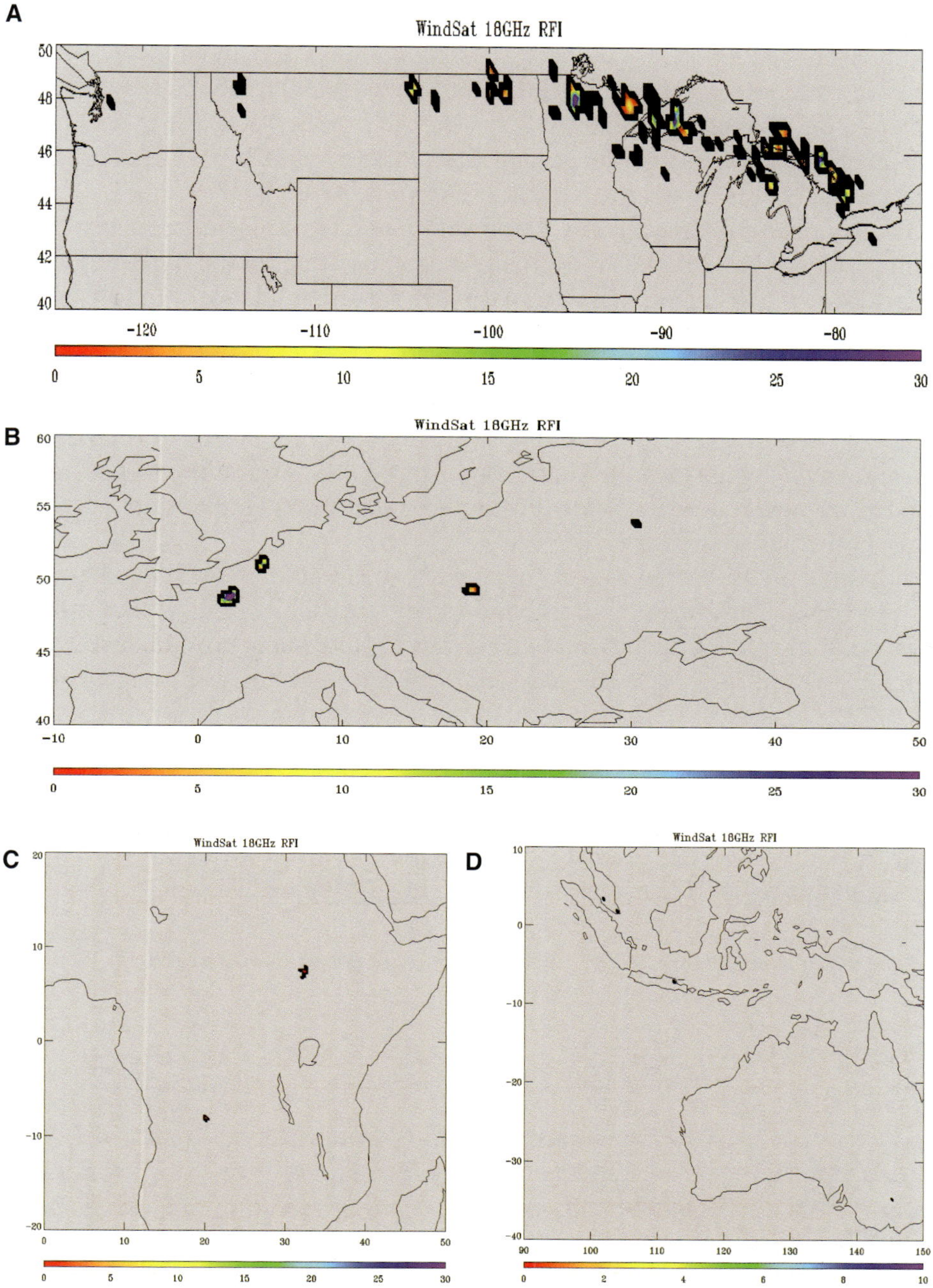

FIGURE 2.20 Cumulative analysis over a 5 year period of WindSat 18.6-18.8 GHz horizontally polarized data indicates sparse occurrences of strong radio frequency interference impacting 18 GHz brightness-temperature measurements over land: (A) North America, (B) Europe, (C) Central Africa, and (D) Southeast Asia/Oceania. Courtesy of the U.S. Naval Research Laboratory.

tive anticollision radar to operate within the bands from 22 to 27 GHz, despite the allocation of the 23.6-24.0 GHz band to the passive services by both the Federal Communications Commission in the United States and ITU globally. Observations at 23.6-24.0 GHz and nearby bands provide the primary data used to estimate atmospheric integrated water vapor, an EDR that drives important atmospheric modes related to severe weather within NWP models (see §2.1).

For a typical five-channel, 22 GHz ground-based upward-looking water vapor profiling radiometer, 1 K of RFI in a channel near the center of the water vapor line can induce a 10 percent error in retrieved water vapor abundance in the lower and mid-level troposphere. This error is comparable to the current performance of such a current technology microwave profiler, and the tolerable RFI level is therefore about 0.1 K. The tolerable RFI level near 31 GHz for total integrated (as opposed to profiles of) water vapor/cloud liquid measurements within the midlatitude coastal environment is about 0.6 K on humid days. Higher RFI levels of up to 1 K can be tolerated for observations of integrated liquid water in clouds and rain where the atmospheric signals are higher.

To date, only little evidence of the impact of RFI at 23.6-24.0 GHz has been documented, partly because automobile radars are still quite new and not yet widespread. In spite of the nascent state of automotive radar, ground-based measurements within 23.6-24.0 GHz have shown the presence of such transmissions. This topic is discussed in further detail in the subsection below entitled "Potential Future Radio Frequency Interference and Its Impact on EESS Observations."

Finding: The rules for out-of-band and spurious emissions in the primary allocated Earth Exploration-Satellite Service (EESS) bands (e.g., 1400-1427 MHz) do not provide adequate interference protection for EESS purposes.

The rules that pertain to the finding above are given in Appendix D.

Unprotected Band

C-Band (6.2-7.5 GHz)

Current space-based observations within C-band, specifically near 6.8 GHz, are used to measure global sea surface temperature and soil moisture. In addition, airborne observations in C-band are used for high-resolution SM mapping for research purposes. Data from flood-prone areas in Texas in 2007 have suggested that airborne mapping at C-band may also be useful for flood forecasting in disaster management. Because there is no EESS allocation within C-band and this portion of the spectrum is heavily used by the Fixed Service, brightness-temperature measurements at C-band over land are currently considered observations of opportunity. The observed area can contain many sources of RFI that require mitigation in order for the data to be useful. Simulated data based on current active spectrum usage have shown that frequency diversity can facilitate effective RFI mitigation in this spectral region. The careful design of receivers and retrieval algorithms can also help facilitate mitigation, but mitigation techniques applied to data from current space-based radiometers are limited in their effectiveness.

The NASA AMSR-E and the WindSat spaceborne radiometers have shown clear evidence of active use impacting C-band EESS measurements (see Figure 2.21) over large portions of global land area. However, the SMMR C-band channel that operated from June 1978 to August 1987 showed little to no evidence of transmissions over North America (Figure 2.22) in this EESS band of opportunity. While the precise bands for these three instruments differ slightly, it has also been qualitatively observed in repeated airborne observations over central Oklahoma in 1999 and 2006 using the same instrument (the PSR/C airborne scanning radiometer) that obvious instances of RFI have tended to increase over time. The major increase in the active usage of C-band spectrum occurring from 1987 to 2003 has reduced the ability to perform EESS observations of opportunity over land. C-band measurements from AMSR-E and WindSat currently provide critical SST products over ocean sufficiently far from the coasts. Ongoing improvements in maritime product accuracies, particularly in near-shore sea surface temperature measurements improved to 0.1-0.2 K accuracy, may thus become limited in the near future by RFI, even far out at sea.

In the examples given in Figures 2.21 and 2.22, AMSR-E imagery illustrates the prevalence and growth of RFI to EESS at C-band. Shortly after the launch of AMSR-E aboard NASA's Aqua satellite in May 2002, it was discovered that the 6.9 GHz passes over land (both ascending and descending and in both V and H polarizations) exhibited anomalous brightness-temperature (T_B) "hot-spots" exceeding 310-320 K that were clearly unrelated to natural surface emission. T_B values also appeared elevated by several degrees over large areas relative to expected values. The RFI not only biased the soil moisture retrievals toward dryness, but caused the multiple-channel iterative algorithm used at launch to fail frequently. Several orbits of data were analyzed, focusing on the United States where the problem appeared to be worst, to see if a simple brightness-temperature index could be devised to detect RFI so that contaminated observations could be ignored. It was found that a simple RFI index could identify the major RFI locations, but low-level RFI covered very large areas and could not be unambiguously distinguished from natural geophysical signals. The AMSR-E RFI was later analyzed globally using a more sophisticated set of indices and statistics.[47] RFI was found at 6.9 GHz over

[47]E.G. Njoku, P. Ashcroft, T.K. Chan and L. Li, "Global Survey and Statistics of Radio-Frequency Interference in AMSR-E Land Observations," *IEEE Transactions on Geoscience and Remote Sensing*, 43(5): 938-947 (2005).

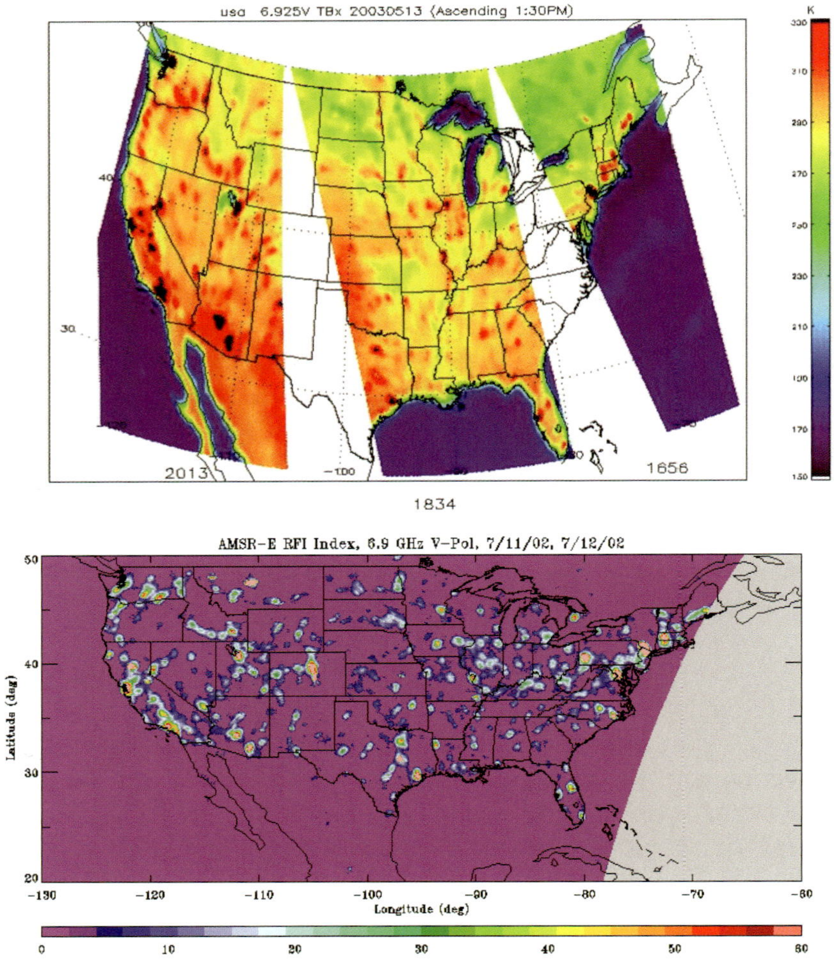

FIGURE 2.21 (Top panel) An example of interference to EESS observations of opportunity at 6.925 GHz primarily from in-band signals arriving via the sidelobes of the main antenna beam of Fixed Service transmitters in legal operation. Passive microwave imagery at 6.9 GHz from AMSR-E on the NASA EOS Aqua platform. The black spots represent high levels of anthropogenic emission that saturate the AMSR-E radiometer primarily over regions of California and Arizona. The red spots over most of the remaining areas of the United States represent contaminated brightness-temperature measurements. (Bottom panel) Radio frequency interference (RFI) is displayed as the perturbation from a zero mean (natural emission) level. Perturbations of up to 50 K are common across the United States, affecting more than 50 percent of the total land area with RFI greater than 5 K. The pervasive nature of the interference makes impossible the retrieval of soil moisture using AMSR-E 6.9 GHz data. SOURCE (top and bottom): L. Li, E. Njoku, E. Im, P. Chang, and K. St. German, "Frequency Interference over the U.S. in Aqua AMSR-E Data," *IEEE Transactions on Geoscience and Remote Sensing,* 42(2): 380-390 (2004), from Figure 8. AMSR-E data are produced by Remote Sensing Systems and sponsored by the NASA Earth Science MEaSUREs DISCOVER Project and the AMSR-E Science Team. Data are available at www.remss.com.

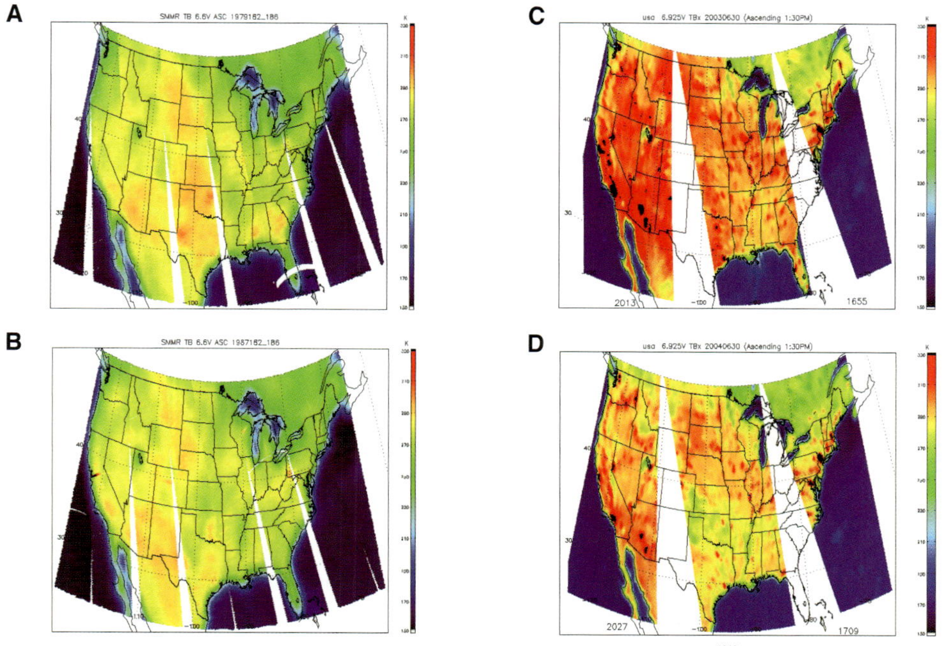

FIGURE 2.22 An example of interference to EESS observations of opportunity at 6.6 GHz. Passive microwave imagery at 6.6 GHz from the Scanning Multi-channel Microwave Radiometer (SMMR) from (A) 1979 and (B) 1987, showing no noticeable brightness temperature from radio frequency interference (RFI). In contrast, passive microwave imagery from the Advanced Microwave Scanning Radiometer-Earth (AMSR-E) on NASA Earth Observing System Aqua from 2003 (C) and 2004 (D) shows substantial RFI. The black spots represent high levels of anthropogenic emission that saturate the AMSR-E radiometer, primarily over regions of California and Arizona. The red spots over most of the remaining areas of the United States represent contaminated brightness-temperature measurements. AMSR-E data are produced by Remote Sensing Systems and sponsored by the NASA Earth Science MEaSUREs DISCOVER Project and the AMSR-E Science Team. Data are available at www.remss.com.

large parts of the Middle East, Asia, and Japan, and even sophisticated statistical procedures could not adequately distinguish RFI from the background of natural brightness variability, nor filter it out in post-processing of the data.[48] Because the 6.9 GHz RFI was so prevalent and difficult to identify and mitigate over the United States, this instrument channel was subsequently ignored in the global AMSR-E algorithm used for the production processing and data archiving of SM data. Reliance was instead placed on the higher-frequency AMSR-E channels that are less sensitive to SM. Over those parts of Europe and Japan where the 10.7 GHz channels were also affected by RFI, no AMSR-E soil moisture retrievals at all were possible.

[48] Ibid.

On a research basis (separate from the global production algorithm), it is still possible to use the 6.9 GHz brightness data for soil moisture retrieval over significant RFI-free global areas such as most of Africa, South America, and Australia.

Extensive analysis of AMSR-E and WindSat data provides a clear picture and plausible explanation for RFI at C-band, but not in other parts of the spectrum. Other RFI surveys have been inconclusive, tied to a single location, and/or have not been able to provide much insight regarding the global status of potential RFI to EESS. The duty cycle, waveforms, emitter spatial distribution, transmitter power, and spectral utilization of the RFI need to be measured to effectively and optimally design RFI mitigation strategies into EESS radiometer systems and to further develop equitable spectrum usage policies. [49] In short, inadequate data on spectrum usage exist. The Federal Communications Commission's (FCC's) 2002 Spectrum Policy Task Force came to this same conclusion:

> More information, however, is needed in order to quantify and characterize spectrum usage more accurately so that the Commission can adopt spectrum policies that take advantage of these spectrum white spaces. Currently, no federal agency or other organization systematically measures temporal spectrum use.[50]

Finding: Better utilization of the spectrum and reduced radio frequency interference for scientific as well as commercial applications are possible with better knowledge of actual spectrum usage.

Progress toward the goal of improved spectrum usage could be made by gathering more information through improved and continuous spectral monitoring. Such monitoring would benefit both the scientific community and commercial interests by allowing more efficient use of the spectrum for communications.

Interference mitigation at C-band has been demonstrated on a limited basis and for particularly strong (and therefore relatively obvious) interference in airborne images of thermal emission at C-band.[51] The radiometer and algorithm were designed to detect spectral variations that were not of natural origin by fitting the spectrum to a standard model, then rejecting channels that compromised the fit

[49]J.R. Piepmeier, "Radio Frequency Survey of the 21-cm Wavelength (1.4 GHz) Allocation for Passive Microwave Observing," in *Proceedings of the 2003 International Geoscience and Remote Sensing Symposium (IGARSS)*, Toulouse, France, 2003, pp. 1739-1741; and presentation by Dennis Roberson, Illinois Institute of Technology, to the committee on September 29, 2007, in Irvine, California.

[50]Federal Communications Commission, *Report of the Spectrum Policy Task Force,* November 2002, p. 10.

[51]A.J. Gasiewski, M. Klein, A.Yevgrafov, and V. Leuskiy, "Interference Mitigation in Passive Microwave Radiometry," *Proceedings of the 2002 International Geoscience and Remote Sensing Symposium (IGARSS),* Toronto, Ontario, Canada, June 24-28, 2002.

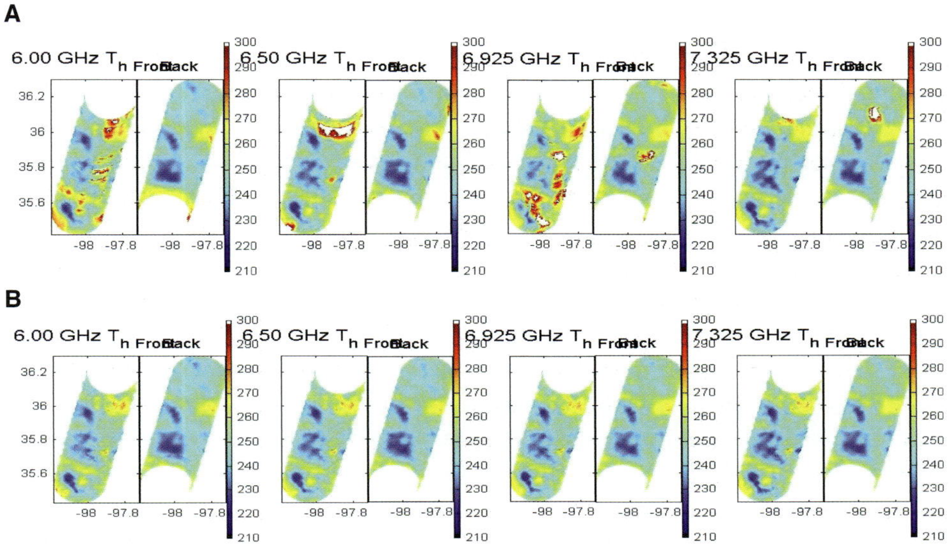

FIGURE 2.23 Polarimetric Scanning Radiometer C-band maps from a swath segment observed during SP99 on July 14, 1999, over central Oklahoma: (A) raw calibrated brightness maps for front and back looks for four subbands and (B) interference-corrected maps using a spectral sub-band algorithm (A.J. Gasiewski, M. Klein, A.Yevgrafov, and V. Leuskiy, "Interference Mitigation in Passive Microwave Radiometry," *Proceedings of the 2002 International Geoscience and Remote Sensing Symposium [IGARSS],* Toronto, Ontario, Canada, June 24-28, 2002). AMSR-E data are produced by Remote Sensing Systems and sponsored by the NASA Earth Science MEaSUREs DISCOVER Project and the AMSR-E Science Team. Data are available at www.remss.com.

to this natural model. The techniques have proven effective at mitigating large-amplitude interference (Figure 2.23). However, they provide no guarantee that interference of amplitudes on the order of the system noise level can be detected and mitigated.

Finding: There is currently inadequate protected spectrum in C-band and X-band for operational passive microwave observations of sea surface temperature, soil moisture, and ocean surface wind speed and direction.

Finding: While unilateral radio frequency interference mitigation techniques are a potentially valuable means of facilitating spectrum sharing, they are not a substitute for primary allocated passive spectrum and the enforcement of regulations.

Finding: Important scientific inquiry and applications enabled by the EESS are significantly impeded or precluded by radio frequency interference (RFI). Such RFI has reduced the societal and scientific return of EESS observatories and necessitates costly interference mitigation, which is often insufficient to prevent RFI damage.

Potential Future Radio Frequency Interference and Its Impact on EESS Observations

Ultrawideband Devices and Anticollision Radar (1-24 GHz)

A major concern for future EESS observations is the proliferation of ultrawideband (UWB) devices that radiate over wide bandwidths at low power, typically in the 2-10 GHz and 22-27 GHz ranges. Automotive collision-avoidance radars that employ the entire 22-27 GHz range have recently been included on new vehicles and are becoming widespread. In particular, the FCC's 2002 approval of the use of UWB devices in the 3-10.6 GHz band and of anticollision radar operation as Part 15 devices near 24 GHz has alarmed the EESS community.[52,53] These sources produce broadband signals that resemble thermal noise, making them difficult to distinguish from natural emissions. The potential for large-scale market penetration of such devices further exacerbates the problem, particularly if they are permitted to radiate across protected frequency bands (particularly in the protected 1.400-1.427 GHz and 23.6-24.0 GHz bands). Emissions from UWB sources in these protected spectral bands present a serious problem, and action will need to be taken to prevent such emissions and limit the numbers of such devices.

Scenarios involving RFI to EESS systems from multiple low-level emitters within the passband and footprint of EESS measurements must be analyzed on a cumulative basis as outlined in Appendix C. In these scenarios the maximum output power of each transmitter and their number per square kilometer are critical factors in EESS compatibility studies. Examples include UWB at 6 GHz and point-to-point transmitters near 57 GHz (see V-band scenarios later in this section).

A study analyzing the impact of losing the protected 23.6-24.0 GHz channel suggests that although the ideal level of RFI in the band is zero, 0.03 K might be established as its maximum permissible value, which is equivalent to –126.84 dBm of RFI within a 500 MHz band.[54] More serious is the fact that unless the RFI level is 10 K or more, the NWP applications cannot reliably flag the data as

[52]See the Glossary in this report for a definition of a Part 15 device.

[53]FCC Press Release, "New Public Safety Applications and Broadband Internet Access Among Uses Envisioned by FCC Authorization of Ultra-Wideband Technology," February 12, 2002, available at http://www.fcc.gov/Bureaus/Engineering_Technology/News_Releases/2002/nret0203.html; accessed January 7, 2010.

[54]S. English, *Assessment of the Requirement for 23.6-24.0 GHz Observations for Weather Forecasting,* Forecasting Research Technical Report No. 440, Exeter, U.K.: Met Office, 2006.

erroneous, thereby degrading forecasts within and downstream of any regions where intermediate-level RFI is present. Such intermediate-level interference is difficult to detect with any confidence except in locations where its effects become extreme. Since automobiles are nearly ubiquitous over land and especially within populated regions where forecasts have the greatest economic value, the problem is endemic to users who rely the most on forecast data. This final point is sufficient to support the exclusion of all intended emissions near the protected EESS band, consistent with the intent of the original regulation.

In addition, there is great concern for the future of EESS measurements of opportunity at C-band. This band covers much of the spectral region commonly used by EESS for measurements of sea surface temperature and soil moisture on an as-available basis. These measurements are critical for accurate weather forecasting, severe weather prediction, and drought prediction, among other applications. The wide proliferation of low-level UWB devices within C-band is a significant concern of the EESS operational and scientific communities (see Appendix C for the density of interferers analysis). Since RFI in EESS operations is cumulative, there is no protection from the impact of a high density of low-level emitters resulting from the strong market penetration of unlicensed products. In these scenarios, all mitigation techniques for AMSR-E and WindSat data would be rendered useless, and important future C-band observations would not be possible without mandatory bilateral mitigation strategies (as described in Chapter 4 of this report).

It is instructive to contrast the scenarios at C-band for EESS, where a large number of emitters contribute to RFI within a single pixel of AMSR-E and WindSat data (especially over populated areas), with the RFI scenario outlined in Appendix D. In the latter case, the impact of RFI on EESS measurements from one or more radars is considered. For cases where only a few high-level emitters in adjacent bands are present (for example, in L-band radar RFI), the measured brightness temperatures are increased by spurious and/or OOB emissions. Such emissions contribute directly to the maximum allowed in-band emissions for EESS; however, the RFI is the result of a single emitter rather than the cumulative effect of many in-band emitters. Although current regulations—if enforced—could preclude the effects of cumulative in-band emissions on EESS systems operating in allocated bands (e.g. 1.400-1.427 GHz and 10.6-10.7 GHz), they are largely ineffective in their present form in limiting OOB and spurious emissions. In considering these scenarios, it should be noted that the present specifications on OOB and spurious emissions were established decades ago, before heavy use was made of bands adjacent to where critical EESS measurements are now conducted and prior to major advances in microwave signal processing and filtering technology. Considerations of new technologies must be made in reassessing the effects of and in regulating OOB and spurious emissions.

Ground-Based Atmospheric Sounding (23.8 GHz, 31.5 GHz, 50-60 GHz, 89 GHz, 183 GHz)

Ground-based microwave radiometers are being used increasingly for the temperature, humidity, and cloud liquid profiles in the lower troposphere for both nowcasting and forecasting. Thus, they are being incorporated into weather observing networks as a replacement and augmentation of the global radiosonde network. It is expected that RMS instrument errors in oxygen-band temperature profiling radiometer measurements will be as low as 0.2 K (or lower) in the future and that the nominal tolerable RFI level for these systems will be 0.02 K.

For a typical five-channel 22 GHz to 30 GHz upward-looking water vapor profiling radiometer, 1 K of RFI in a channel near the center of the 22.235 GHz water vapor line can induce a 10 percent error in retrieved water vapor abundance in the lower and mid-level troposphere. This error is comparable to the current performance of such a profiler, and the tolerable RFI level is therefore about 0.1 K. It is expected, however, that the absolute accuracy of ground-based systems will increase as the models and instruments improve, possibly attaining an absolute accuracy of 0.2 mm of precipitable water vapor (PWV). Since each millimeter of PWV produces approximately 1.4 K of signal at 23.8 GHz, RFI must be less than 0.03 K, assuming a maximum tolerable interference of 10 percent of the sensitivity of the instrument. Higher RFI levels of up to 1 K can be tolerated for observations of integrated liquid water in clouds and rain.

Wideband anticollision radars are being licensed and produced in the 22-26 GHz region of the 22-30 GHz wave band, which spans the radio astronomy reserved quiet band at 23.6-24 GHz. These active sources are difficult to discriminate from thermal noise, even with elegant and costly detection methods, and are expected to be an ever-increasing problem to ground-based water vapor (humidity) profiling.

Ground-based radiometers receiving around 89 GHz are important in that they are used to discriminate between cloud liquid water and ice. The transitions between the ice-liquid-vapor phases of water drive the thermodynamic energy transport cycles of the atmosphere and are therefore important for monitoring and predicting weather. Knowledge of these three phases is also critical to understanding planetary albedo and planetary radiative transfer, and therefore climate change and global warming, as well. There is a protected primary radio astronomy band at 86-92 GHz, but as mentioned elsewhere in this report, it is difficult to enforce against intrusions by spurious and out-of-band transmissions. Active technologies up to 110 GHz are being developed, in part due to military interest in and funding for active radars around 94 GHz. The growing availability of these high-frequency technologies in this wave band will undoubtedly result in problems from RFI for EESS observations.

The strong water vapor line centered at 183 GHz is observed for water vapor profiling in dry climates such as high-altitude astronomical observatories and arctic and desert regions. Because of the level of technology required at these high frequencies, little interference in this region is foreseen in the near future.

Other Concerns

SST Measurements at C-Band and X-Band (5-10 GHz)

Of particular future concern is RFI affecting continuous all-weather microwave sea surface temperature measurements in littoral regions that are critical for severe storm forecasting and weather and climate studies (see Figure 2.8). These measurements rely principally on observations at 5-10 GHz, which are generally sensitive to surface temperature changes while being insensitive to clouds. Active services using spectrum adjacent to and within the EESS allocation at 10.6-10.7 GHz can make SST measurements difficult or impossible at this band. UWB devices that radiate in the 2-10 GHz range could be particularly problematic in the future. It is also important to note that 10.6-10.68 GHz is shared with the Fixed Service, and in several areas worldwide, significant interference has been measured and continues to increase. Several EESS satellites have improved on TMI's 10 GHz measurements of SST by including observations of C-band microwave brightness temperatures, typically near 6.8 GHz. These measurements specifically improve the accuracy of all-weather SST measurements in cold regions and are less prone to being affected by heavy clouds and precipitation. However, uncontaminated measurements of environmental parameters near 6 GHz are becoming more difficult to obtain owing to the high usage of the C-band spectrum and the lack of any EESS allocation adequate to support SST measurements. While the problem of contamination of 5-10 GHz SST measurements exists over all of the global oceans, it is particularly an issue in littoral regions where severe weather is economically important and population density (including ship traffic) is high (see also §2.1).

V-Band (50-64 GHz)

A number of currently operating space-based instruments use the atmospheric oxygen absorption band (50-64 GHz) to estimate profiles of atmospheric temperature and moisture. These measurements are central to NWP, severe weather forecasting, and climate analysis. International frequency allocations provide a shared "primary" status to EESS in the 57.0-59.3 GHz range, and these frequencies are currently used by several space-based radiometers, including the Advanced Microwave Sounding Unit and the Special Sensor Microwave Imager/Sounder. Both of these sensors operate on multiple satellites to provide full global coverage every few hours (see Table 2.2). AMSU sensors operating in the 50-59 GHz band may be the single most important data source enabling useful global weather forecasts up to 7 days in advance.

In response to a growing interest in the active use of this part of the spectrum, EESS scientists have begun analyzing the potential for future interference to remote sensing measurements at V-band. The wide bandwidth available and small device sizes that can be manufactured make this potentially fertile ground for commercial interests.[55] A recent FCC notice of public rule making (NPRM) requested an allowance for increased power emission levels for sources operating within 57-64 GHz, which includes the ITU-protected 57.0-59.3 GHz portion used for weather-related sensing by many satellites and weather forecasting services.[56] Unfortunately, the FCC NPRM included no analysis of the potential impact of these increased power levels on essential EESS passive measurements from AMSU or related instruments, even though it is currently envisioned that wireless systems operating near 60 GHz will become ubiquitous consumer devices for applications such as local DVD broadcasts and personal networking. While atmospheric absorption limits the range of active users' transmissions, attenuation from the surface to the top of the atmosphere is not complete (as shown in Figure 1.2). A sufficiently high spatial density of low-power emitters on the ground can affect spaceborne microwave observations. Members of the EESS passive community raised this issue in comments filed in response to the FCC's NPRM, and the FCC's decision is still forthcoming as of the time of this writing.[57] The community is also interacting with IEEE standards organizations to determine the possible impact of such wireless systems on future EESS observations.[58]

It is clear that RFI degradation of EESS measurements and weather forecasting services appears to be likely if widespread unlicensed transmissions in these bands begin. Consideration should be given to limiting the strength and density of transmitters in this band (see Appendix C) in order to address the concerns of EESS. It may well be that no practical limit exists if such devices are sold as unlicensed

[55]B. Bosco, "Emerging Commercial Applications Using the 60 GHz Band," IEEE Wireless and Microwave Technology conference (WAMICON) 2006, proceedings; B. Razavi, "Gadgets Gab at 60 GHz," *IEEE Spectrum*, February 2008.

[56]*In the Matter of Revision of the Commission's Rules Regarding Operation in the 57-64 GHz Band*, Notice of Proposed Rulemaking, 22 FCC Rcd 10505 (2007).

[57]IEEE Geoscience and Remote Sensing Society, "Comments to the proposed revision of the Commission's Rules Regarding Operation in the 57-64 GHz Band,"available at http://fjallfoss.fcc.gov/prod/ecfs/retrieve.cgi?native_or_pdf=pdf&id_document=6519741794; accessed June 9, 2009.

[58]It is noted that while considerable resources are often available to be applied toward legal filings by active users of the spectrum, the nongovernmental scientific community has had little or no financial support for pursuing such legal matters. Virtually all responses from the nongovernmental EESS and RAS communities to NPRMs are the result of either voluntary efforts (in the case of university personnel) or are in direct reaction to threats to the viability of the passive services (in the case of industry personnel).

and thus potentially used without limit. However, there is no apparent technical reason why the wider band 59.3-64 GHz could not alternatively satisfy essentially all commercial requirements for ubiquitous devices since such bandwidths in a single device far exceed the capacities of most home fiber and cable systems that offer hundreds of television channels and other services.

High Frequencies (>100 GHz)

In order to improve the understanding of the chemistry associated with stratospheric ozone depletion, it is necessary to observe the global distributions of a wide array of trace gases.[59] Measurements are made by observing narrow spectral line emissions. The frequency requirements of those measurements are dictated by molecular quantum transitions of the gases under consideration. Trace gases of particular interest include ozone, chlorine, hydrogen, bromine, and water vapor. NASA's Microwave Limb Sounder (MLS) and associated follow-on instruments have been designed for trace gas observations.[60] The EOS version of MLS operates in five primary spectral bands near 118, 190, 240, 640, and 2500 GHz.[61] The specific passbands and minimum detectable signals for MLS are listed in Table 2.3. RFI should be kept at or below one-tenth of the minimum detectable signals levels noted in the table. While no RFI has been reported to date, it is envisioned that the bands above 100 GHz may become commercially useful to the active services in the coming decades.

In the near term, the Submillimeter Infrared Radiometer Ice Cloud Experiment (SIRICE) mission is being designed to measure cloud ice water path using passive channels above 100 GHz. SIRICE is currently in pre-Phase A development at NASA. Design studies have identified three channels (including frequencies, bandwidths, and rms measurement errors) for SIRICE required to retrieve IWP with the necessary accuracy and precision. The spectral requirements are summarized in Table 2.4. RFI contamination of SIRICE observations should be at or below one-tenth of the NEΔT levels noted in the table if the scientific integrity of the IWP retrievals is to be maintained.

[59]S. Solomon, "Stratospheric Ozone Depletion: A Review of Concepts and History," *Reviews of Geophysics,* 37(3): 275–316 (1999).

[60]J.W. Waters, W.G. Read, L. Froidevaux, and R.F Jarnot, "The UARS and EOS Microwave Limb Sounder (MLS) Experiments," *Journal of Atmospheric Science,* 56: 194-217 (1999).

[61]J.W. Waters et al., "The Earth Observing System Microwave Limb Sounder (EOS MLS) on the Aura Satellite," *IEEE Transactions on Geoscience and Remote Sensing,* 44(5): 1075-1092 (2006).

TABLE 2.3 EOS Microwave Limb Sounder Instrument Spectral Coverage and Sensitivity for Measurement of Trace Gases in the Upper Atmosphere

Passband (GHz)	Minimum Detectable Signal (K)
115.3-122.0	0.1
177.2-206.2	0.03
221.4-240.5	0.1
606.7-657.5	0.1
2481.9-2506.0	0.1

SOURCE: J. Waters, R.E. Cofield, M.J. Filipiak, D.A. Flower, N.J. Livesey, G.L. Manney, H.C. Pumphrey, M.L. Santee, P.H. Siegel, and D.L. Wu, "An Overview of the EOS MLS Experiment," NASA EOS MLS DRL 601 (part 1), ATBD-MLS-01, JPL D-15745/CL#04-2323, ver. 2.0, January 7, 2005.

TABLE 2.4 Submillimeter Infrared Radiometer Ice Cloud Experiment (SIRICE) Instrument Spectral Coverage and Sensitivity Requirements for Measurement of Ice Water Path

Center Frequency ± Double Sideband Offset (GHz)	Bandwidth (GHz)	NEΔT (K)	Polarization
183.31±1.5	1.4	0.7	Vertical
183.31±3.5	2.0	0.6	Vertical
183.31±7.0	3.0	0.5	Vertical
325.15±1.5	1.6	1.8	Vertical
325.15±3.5	2.4	1.4	Vertical
325.15±9.5	3.0	1.3	Vertical
448.00±1.4	1.2	2.3	Vertical
448.00±3.0	2.0	1.8	Vertical
448.00±7.2	3.0	1.5	Vertical
642.90±6.7	2.8	1.9	Vertical
642.90±6.7	2.8	1.9	Horizontal
874.40±4.5	6.0	1.9	Vertical

2.6 SUMMARY OF THE IMPORTANCE OF AND RISKS TO CONTINUED CONTRIBUTIONS OF THE EARTH EXPLORATION-SATELLITE SERVICE IN THE FUTURE

The Earth Exploration-Satellite Service (EESS) provides critical and unique measurements that support (1) day-to-day weather and other environmental operations, (2) climate research, and (3) model development and other scientific advances in Earth observation. EESS measurements are currently impacted by RFI at all key frequencies up to 19 GHz, and likely at 24 GHz and higher frequencies soon. There is also potential for significant future interference to EESS systems operating at 50-60 GHz. This interference occurs whether the band of concern is assigned to the passive services exclusively, shared with other services, or not assigned to EESS but has unique physical properties that demand observation when interference is absent. Unless these issues are addressed in a timely manner, the effectiveness and utility of EESS will likely be increasingly compromised, particularly as wireless services and unlicensed devices proliferate. Most problematic are future ubiquitous unlicensed ultrawideband consumer devices that can proliferate without limit.

Box 2.3 illustrates a sporadic record of achievement in appropriately allocating spectrum and/or coordinating technology development between EESS and competing active services. A technology advisory body, incorporating members from all relevant services, could help mitigate such failures. Such an entity would link EESS and other relevant active and passive communities in an early identification of issues and opportunities regarding competing spectral needs and shared standards development. Such a holistic body would supplement the more adversarial and segmented bodies that currently provide most such advice.

BOX 2.3
Illustrative Examples of Successes and Failures in Frequency Coordination That Affect the Earth Exploration-Satellite Service (EESS)

Successes

- European and Japanese transition to 77 GHz band for automobile radar, avoiding 23-24 GHz.
- The development of airborne sub-band-based radio frequency interference (RFI) mitigation methods that delete single strong interference signals, although not weak or diffuse interference.
- The International Telecommunication Union trade-off of allocations to obtain stronger protection at more important bands at 50-57 GHz.
- The migration of new instrument specifications toward protected bands (Advanced Technology Microwave Sounder, Special Sensor Microwave/Imager, Special Sensor Microwave/Imager Sounder, Conical Microwave Imager Sounder, and Microwave Imager/Sounder).

Failures

- The lack of engagement between the auto radar community, Earth Exploration-Satellite Service (EESS), and regulators during the technology's early development.
- The lack of accepted remedies when unlicensed devices producing limited EESS interference multiply in numbers so as to collectively damage EESS and other services.
- The lack of global exclusive EESS allocations at 18.7 and 10.65 GHz; critical bands experiencing RFI.
- No allocation of a protected band at C-band.
- The difficulty in effectively employing lower-frequency bands (e.g., 1400-1427 MHz) owing to RFI; apparent inadequate protection for EESS operation in the exclusively passive 1400-1427 MHz band.

3

The Radio Astronomy Service

Over the past 75 years, astronomical observations at radio frequencies have transformed the understanding of the universe. They have allowed fundamental scientific issues to be addressed—these include the creation and ultimate fate of the universe, the distribution of matter and primordial energy in the universe, and the environment and manner in which stars and planets form. Many astronomical discoveries that have captured the imagination of astronomers and the public alike were made accidentally with radio telescopes; a list of such discoveries would include that of the primordial cosmic microwave background (CMB), celestial masers, and pulsars—the latter being the dense, fast-rotating, radio-emitting remnants of massive stars. With powerful new facilities such as the Atacama Large Millimeter Array (ALMA), the potential for unexpected discoveries will grow substantially. As was fittingly said in this context years ago by two famous radio astronomy pioneers, "We cannot discuss plans to discover the unsuspected . . . ,"[1] but the parade of new, unexpected discoveries has been continuous since the beginning of radio astronomy in the 1930s. With the unprecedented regimes of sensitivity that will arrive with new and planned instruments, one can expect that further remarkable discoveries will be made.

Astronomical discoveries have been made possible by the steady and enormous improvement in sensitivity that is shown in Figure 3.1. In this graph the ordinate represents sensitivity and is on a logarithmic scale; there has been an improvement

[1]Pawsey and Bracewell, *Radio Astronomy*, Oxford, United Kingdom: Clarendon Press, 1955, p. 296.

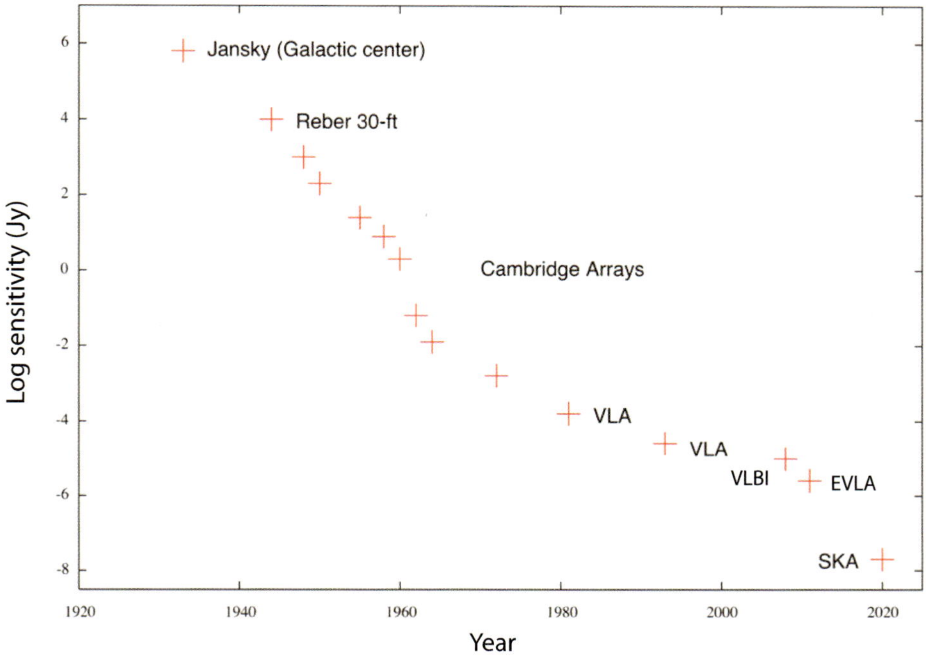

FIGURE 3.1 The minimum detected or detectable signal in flux density versus the year of measurement. The sensitivity is proportional to the temperature of the receiver system and inversely proportional to the collecting area and the square root of both bandwidth and integration time. For measurements after year 1990, an integration time of 12 hours is assumed. The rapid improvement over time is due to system improvements, including the decrease in system temperature (solid-state technology), the increase in collecting area (cost and construction efficiency), and the increase in bandwidth and integration time (electronic and digital technology). The improvement from 1933 to 1983 is about 10 orders of magnitude, a halving time of less than 2 years: a performance improvement similar to that described by Moore's law. Acronyms in the figure are defined in Appendix F. Figure adapted and updated from J.M. Moran, "Peter Mezger and the Development of Radio Astronomy in the U.S. and Germany, and the Discovery of Radio Recombination Lines," pp. 475-488 in *The Nuclei of Normal Galaxies, Lessons from the Galactic Center, Proceedings of the NATO Advanced Research Workshop,* NATO Advanced Science Institutes Series C, Vol. 445, A. Harris and R. Genzel, eds., Kluwer, Dordrecht (1994).

of 10 billion in 70 years, and there will be another improvement by a factor of 1,000 from the Very Large Array (VLA) to the Square Kilometer Array (SKA) when it is built. (See Table 3.1 in §3.2, "Radio Observatories and Radio Telescopes," for the characteristics of the newer instruments.)

The current scientific questions that are motivating the construction of these new telescopes are no less exciting than those that were resolved in the past. Obvious

BOX 3.1
Nobel Prizes in Physics for Developments and Discoveries in Radio Astronomy

Radio astronomy has been internationally recognized for its fundamental contributions to knowledge. The Nobel Prize in physics has been awarded to eight radio astronomers. The names of the scientists who led the teams to these discoveries, the year of award, and a brief description of the prize-winning science, are listed below.

- *Sir Martin Ryle, 1974,* for the development of aperture synthesis, and *Antony Hewish, 1974,* for the discovery of pulsars;
- *Arno A. Penzias and Robert W. Wilson, 1978,* for the discovery of the cosmic microwave background radiation;
- *Russell A. Hulse and Joseph H. Taylor, Jr., 1993,* for establishing the emission of gravitational waves by close binary pulsar systems, as predicted by general relativity; and
- *John C. Mather and George F. Smoot, 2006,* for demonstrating that the cosmic microwave background radiation has a blackbody spectrum and for discovering spatial fluctuations in the radiation.

examples include the exploration of planetary systems in formation around other stars, measurements of neutral hydrogen in the early universe, and the study of star formation in distant galaxies. Furthermore, it is through radio observations that the discovery of life-indicating molecules in other planetary systems might be made.

The scientific and technical advances of radio astronomy have been internationally recognized, as listed in Box 3.1. The Nobel Prize in physics has been awarded to eight radio astronomers in the past 40 years.

3.1 THE SCIENTIFIC IMPACT OF RADIO ASTRONOMY

What follows is a summary of the scientific advances made possible in a few areas by radio astronomy. A discussion of some advances expected in the near future is also provided.

Origin of Planets and the Solar System

Speculations concerning the origin of the solar system stretch far back in the science and philosophy of humans. During the coming decade, the capability of understanding the origins and evolution of other planetary systems and thereby coming to understand the origin of our own planetary system will exist: ALMA and the Expanded Very Large Array (EVLA), both coming online in a few years, will make it possible to detect planets in formation around other stars. ALMA and EVLA will enable the study of the structure, dynamics, and temperature of the material from which planets are forming. The planned Square Kilometer Array will enable detailed studies of such disks. The key strengths of radio measurements that will enable these studies are their ability to trace the distribution of gas and dust throughout the disk, to study the dynamics and temperature of the material involved in planet formation, and to follow the accretion of material as it develops from the tiny, submicron dust particles characteristic of the interstellar medium to centimeter-sized clumps, the first critical step in the formation of terrestrial planets. These radio capabilities are unique in enabling scientists to learn about the physical and dynamical processes that govern the planet-formation process and its outcome—a planetary system. They will be able to "see" the formation of giant planets through the gravitational and thermal influence of these planets on the surrounding gas. Scientists will see disks with gaps and inner clearing zones that are caused by planets. They will be able to follow the orbits of the planets by how they sculpt the disk and to study characteristics of the planets by probing their interaction with the disk material.

At present, search techniques for extrasolar planets, or exoplanets, are strongly biased toward finding large planets close to their host star; and correspondingly, the 358 planetary systems known as of November 2009 are very different from our own solar system.[2] They typically contain one or more Jupiter-like giant planets in orbits closer than that of Earth, and with eccentricities exceeding those of any planet in our solar system. There is no well-accepted theory for how such planets form or why they should be common. Prior to the discovery of exoplanets, our solar system was thought to be typical, and a template for all planetary systems. This is now known not to be true, and our understanding of the diverse outcomes of formation is significantly incomplete. So that this formation problem can be properly addressed, observations of many young stars are needed. These observations will lay the groundwork for an understanding of the many possible outcomes of the planet-formation process and how terrestrial planets fit into the general picture.

The new knowledge of the existence of other planetary systems gives rise to many intriguing questions. Does life exist elsewhere, or is it unique to the solar system? Could there be a common starting point for life? The abundant and complex chemistry of the interstellar medium and of protoplanetary systems possibly provides an answer. More than 140 molecules have been discovered in the interstellar medium. Those with more than four atoms are dominated by carbon, nitrogen, oxygen, and hydrogen. The 31 molecules with seven atoms or more are nearly all organic molecules. They include glycoaldehyde (a simple sugar), and urea and glycine (the latter being a simple amino acid common to life) may have

[2]Data from http://exoplanets.org/, accessed November 24, 2009.

been detected. Clearly, the carbon-nitrogen-oxygen chemistry that dominates life on Earth also dominates the complex chemistry of space.

The radio spectrum is the place to pursue a connection between astrochemistry and prebiotic terrestrial chemistry because it gives access to the wealth of spectral lines. With the sensitivity and resolution of the coming generation of radio telescopes, it will be possible to search for sugars and amino acids and to follow the flow of chemistry from molecular clouds into protoplanetary systems. Is there a strong interstellar heritage to the chemical compounds that comets and other bodies delivered to early Earth? What is the dominant chemistry of a protoplanetary nebula and how does that change the chemical composition of the planet? Was life on Earth seeded by interstellar molecules?

In addition to these questions, others arise because the molecular composition of interstellar and protoplanetary material is strongly impacted by the physical processes that act on the gas. Selected molecules can act as tracers to follow specific physical processes. For example, silicon monoxide (SiO) is commonly used as a tracer for strong shock waves associated with outflow activity, because silicon is heavily depleted onto dust grains, which are readily destroyed by shocks. That destruction liberates silicon into the gas phase, and this silicon is quickly incorporated into SiO. Methanol is a similar tracer for weak shocks, which evaporate ices. These tracers, and others presumably yet to be discovered, will provide important insights into the processes that shaped our solar system and that shape other planetary systems.

Now that many planetary systems are being discovered, the search for signs of extraterrestrial life is becoming more compelling. The many planets that will be discovered in the "habitable zone" in the coming years are obvious targets. Searching in the radio band is thought to be the optimum strategy, and some limited searches have already been made with the telescope at the Arecibo Observatory and with other smaller telescopes, but no results have yet been achieved. The Allen Telescope Array (ATA), a dedicated instrument for searching for extraterrestrial signals, is completing its first stage of construction as of this writing and will begin work soon. It will be a multibeam telescope, able to look at many stars simultaneously. This search for an extraterrestrial civilization, while a "long shot," is seeking an answer to a basic and profound question: Are we alone in the Galaxy?

Origin and Evolution of the Universe

In the past few decades, cosmology, the study of the origin and evolution of the universe, has been revolutionized. Whereas 30 years ago only a few broad facts in this field were known, today cosmology is a quantitative science with specific, testable hypotheses. This revolution stemmed from advances in astronomical techniques that broadened astronomy from its origin in the optical wave band to the entire electromagnetic spectrum. This expansion across the spectrum was

> **BOX 3.2**
> **Redshift**
>
> The continual expansion of the universe stretches electromagnetic waves so that they are received on Earth at a frequency lower than the frequency that they had when emitted. This effect is known as *redshift,* because light is shifted toward the red end of the spectrum as the distance is increased. Also, because the velocity of light is finite, more-distant galaxies are seen as they had been at earlier times. Looking at distant galaxies, one sees the universe at an early epoch.

pioneered by radio astronomy, which has been essential to the study of cosmology because radio astronomy alone can detect the bulk of the coldest matter in the universe, and can detect it at enormous distances and early times (see Box 3.2). We now know that the observable universe has expanded from its origin in a Big Bang some 14 billion years ago. It cooled as it expanded, and nuclei of hydrogen and helium were formed in dense opaque plasma. With further cooling, nuclei and electrons combined into atoms, and the universe became transparent but now dark, since as yet there were no stars. In subsequent evolution, the higher-density regions were able to collapse under their own gravity, giving rise to the first stars and galaxies (see Figure 3.2).

Fifty years ago the space density of bright radio galaxies was found to increase with distance faster than expected from the expansion, demonstrating the evolution of the universe and revealing a remarkable epoch of galaxy formation some 10 billion years ago. It was through this simple observation that radio astronomy ruled out the rival, steady-state theory of a non-evolving universe and favored evolutionary models in which the universe has expanded from a compact, hot origin.

Radio astronomy also provides the strongest evidence for the Big Bang through the discovery of the cosmic microwave background (CMB) radiation in 1965. This background radiation fills space and has an accurately measured blackbody spectrum with a temperature of 2.725 K and a broad peak at about 100 GHz. This radiation was emitted some 400,000 years after the Big Bang, at a time when the universe had a temperature of about 3000 K and was becoming transparent. Since that time, the radiation has been stretched by a factor of about 1,000 through the expansion of the universe, and the temperature has decreased by the same factor. Because this radiation is so weak and so highly isotropic, it is difficult to distinguish from local sources of noise. Only very careful observations have been able to demonstrate its existence.

The CMB has proved to be a gold mine of information about the early universe. The radiation comes from early times when the universe was nearly homogeneous,

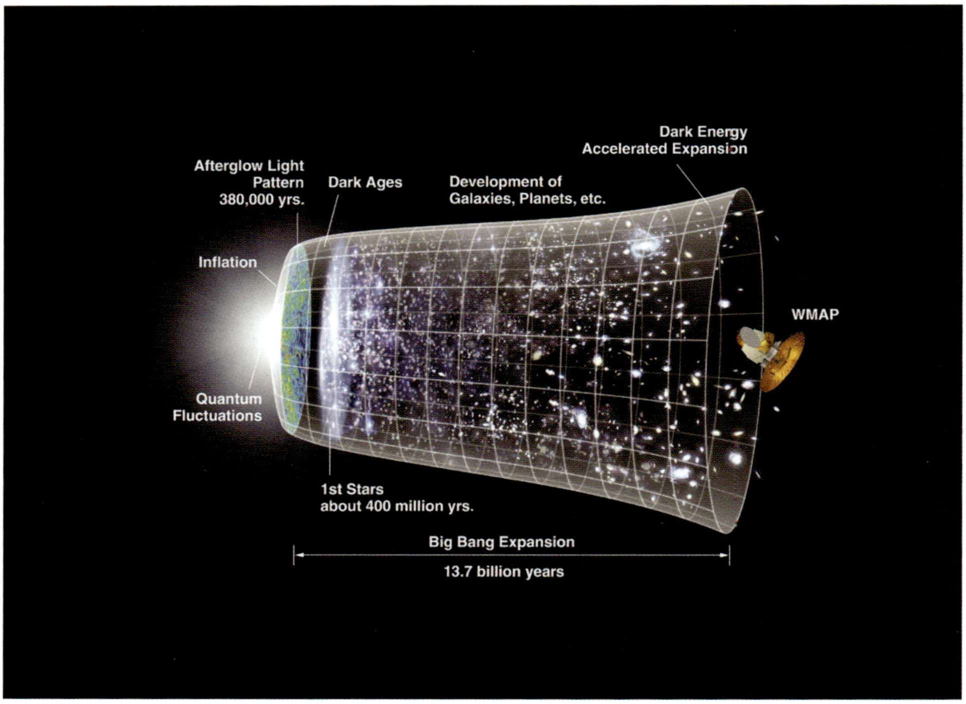

FIGURE 3.2 Artist's conception of the history of the universe. Time runs from left to right. The universe was born in an explosion popularly called the "Big Bang," which perhaps came from a "quantum fluctuation," a phenomenon well known in physics. After a period of hyper-expansion ("inflation"), the universe settled to a nearly steady expansion rate. As the plasma became neutral, the afterglow died out, and the universe became dark. After hundreds of millions of years, gravitational contraction of the material in the original density fluctuations produced the first stars, which gave off light, and so the "Dark Ages" ended. Further generations of stars formed, and galaxies and black holes coalesced from the stars. The universe became more complex and now is evolving rapidly, with many varieties of stars and galaxies and exotic objects, including a planet containing sentient beings who are able to contemplate this vast universe. Results from the Wilkinson Microwave Anisotropy Probe (WMAP) satellite (shown in the figure) were used to make the afterglow pattern. Image courtesy of NASA/WMAP Science Team.

but even then there were small density and temperature fluctuations that became the seeds of stars and galaxies. After extensive searches, the Cosmic Background Explorer (COBE) satellite found these fluctuations in 1992, at a level of 1 part in 100,000 of the background temperature. The fluctuations appear to be random on the sky, but they have a characteristic angular scale of approximately 1 degree, which reveals properties of the plasma from which they were emitted. Measurements of the angular power spectrum of the fluctuations have fixed the conditions of the universe at the emission time, when the plasma changed to a neutral gas of hydrogen and helium. Along with observations in other wave bands, radio observations of the CMB have revealed that most of the material in the universe cannot be "normal matter"; it must be something that does not emit or absorb electromagnetic radiation: "dark matter." In addition, 70 percent of the density is made up of "dark energy," which has a repulsive antigravity effect, causing the expansion of the universe to accelerate.[3]

The fluctuations of the CMB have immense cosmological significance, and they are being studied with many instruments. The emission is broadband but peaks at a few hundred gigahertz, where atmospheric emission is a serious contaminant. Hence, the instruments are located on high mountain sites, on balloons, or on satellites. Very wide bandwidths are needed to detect the tiny signals. The CMB fluctuations are linearly polarized at about the 10 percent level, and this provides further insights into the early universe. CMB studies provide a testing ground for theories of fundamental physics and theories on the nature of space and matter, at energies that cannot be reached by experiments on Earth.

Between the epoch of recombination, when the universe became transparent and the CMB was emitted, and the epoch of galaxy formation, when stars first began to light up the universe, lies the "Dark Ages" of the universe (see Figure 3.2). This period cannot be studied by optical astronomy, but radio provides a window by way of emission from neutral hydrogen. Over the next decade this study will be one of the major thrusts in radio astronomy. The emission, redshifted from 1.4 GHz, will be detected at much lower frequencies, 200 MHz and below. It will be very faint, and radio interference will be a serious concern. Such observations will have to be made from remote sites and will require careful attention to the mitigation of radio frequency interference (RFI).

Pulsars and General Relativity

Pulsars are ultradense collapsed cores of heavy stars in the form of neutron stars that have completed their nuclear burning and exploded. Pulsars have a very strong magnetic field and generate a radio beam that, because the neutron star is spinning, produces radio flashes in the same manner that a lighthouse generates optical flashes. In some cases, the pulsar, remarkably, is spinning at about a thousand times a second, leading to the term *millisecond pulsars*.

Because a pulsar is ultradense, its gravity is ultrastrong, and it provides a natural laboratory for the testing of Einstein's theory of general relativity (GR). One predic-

[3]D.N. Spergel, L. Verde, H.V. Peiris, E. Komatsu, M.R. Nolta, C.L. Bennett, M. Halpern, G. Hinshaw, N. Jarosik, A. Kogut, M. Limon, S.S. Meyer, L. Page, G.S. Tucker, J.L. Weiland, E. Wollack, and E.L. Wright, "First Year Wilkinson Microwave Anisotropy Probe (WMAP) Observations: Determination of Cosmological Parameters," *Astrophysical Journal Supplement*, 148: 175-194 (2003).

tion of GR is that the orbit of a pulsar in a binary stellar system slowly decays due to the emission of gravitational waves (Figure 3.3). The measurements accurately fit the prediction and prove that gravitational waves do exist. For this demonstration, Hulse and Taylor were awarded the Nobel Prize in physics in 1993 (see Box 3.1).

However, this orbital decay is a "weak-field" effect, and GR has not yet been tested in the "strong-field" case. This leaves a fundamental question in physics: Is Einstein's theory the final word in our understanding of gravity? Important questions are unanswered: Can GR correctly describe the ultrastrong field? Are its predictions for black holes correct? Is the cosmos filled with a stochastic gravitational-wave background? Radio observations of pulsars now approach these questions, and the largest radio telescopes, including the Green Bank Telescope (GBT) and the Arecibo Observatory, and especially the SKA, should give some answers. These telescopes offer the possibility of probing the strong-field realm of gravitational physics by finding and timing many pulsars. The ultimate goal is to obtain extremely tight limits on deviations from GR, to a level a thousand times better than present solar-system limits.

In the coming years, radio observations will identify hundreds of millisecond pulsars across the sky. Timed to high precision (~100 ns, the time that it takes light to travel 100 feet), these pulsars will act as multiple arms of a cosmic gravitational-wave detector. This "telescope" will be sensitive to gravity waves at frequencies of nanohertz and will complement the much higher frequencies accessible to direct gravitational-wave detectors such as the Advanced Laser Interferometer Gravitational Wave Observatory (LIGO, ~100 Hz) and the Laser Interferometer Space Antenna (LISA, 1 mHz). The largest radio telescopes will be crucial for these observations.

Galactic Nuclei and Black Holes

The first stars and galaxies formed out of the fluctuations in the early universe. A detailed understanding of how these processes unfolded will probably be one of the major achievements of astronomy in the coming decades. Astronomers have concluded that most galaxies have a giant black hole in their nuclei, with mass between a million and a billion times the mass of the Sun (see Box 3.3). It is not known whether the black holes formed first and galaxies of stars formed around them or the galaxies formed first and the black holes later condensed from the inner core. A remarkable correlation, however, has been found between the mass of black holes in galaxies and the mass of the halo of stars that surrounds them.[4] This relation implies the existence of some regulatory or feedback process linking the black hole and its halo of stars. Over cosmic time, a galaxy grows through mergers

[4]L. Ferrarese and D. Merritt, "A Fundamental Relation between Supermassive Black Holes and Their Host Galaxies," *Astrophysical Journal*, 539(1): L9-L12 (2000).

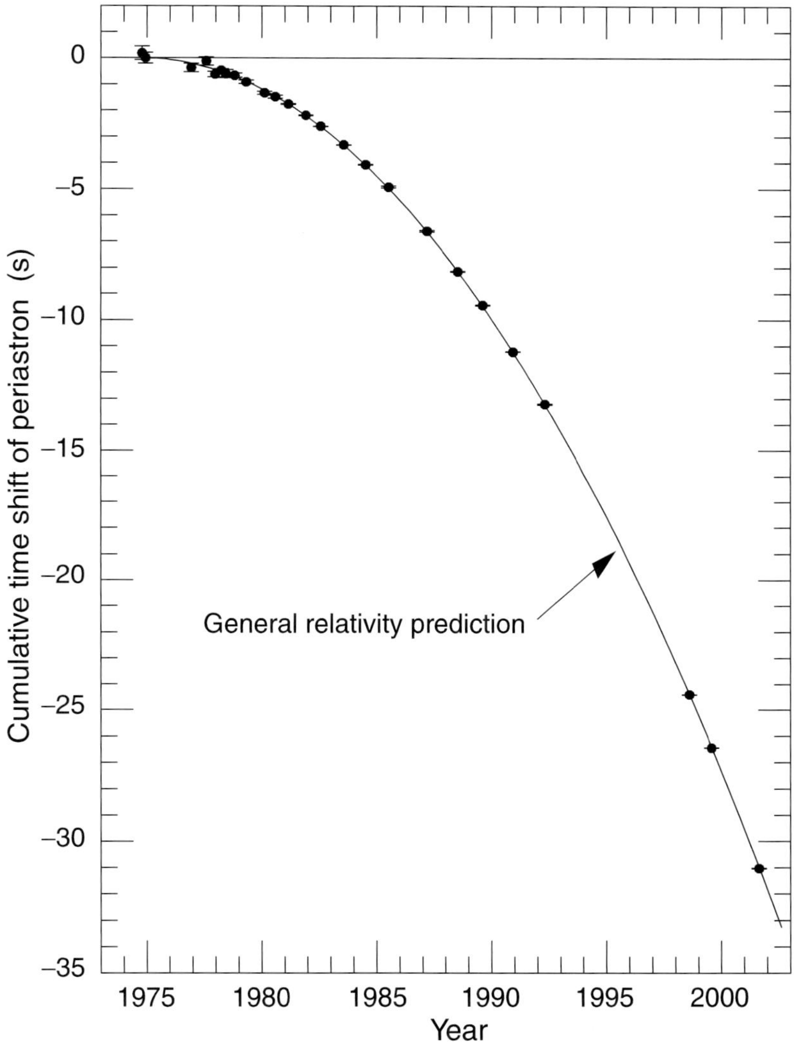

FIGURE 3.3 The results of 30 years of observations at the Arecibo Observatory of the radio-emitting pulsar B1913+16. The pulsar is in orbit around a companion neutron star. General relativity (GR) predicts that the orbits of the two stars will shrink as orbital energy is lost to gravitational radiation. This figure shows the first detection of this effect: measurements of the orbital phase (the data points) exactly match the prediction (solid line) calculated with GR. SOURCE: J.M. Weisberg and J.H. Taylor, "The Relativistic Binary Pulsar B1913+16: Thirty Years of Observations and Analysis," Astronomical Society of the Pacific Conference Series, Vol. 328, F.A. Rasio and I.H. Stairs, eds., 2005.

BOX 3.3
Black Holes

Einstein's theory of gravity (general relativity) predicts that when matter is compressed sufficiently, it contracts into a region of space where gravity is so strong that nothing, not even light waves, can escape. Hence, this ultimate compression forms a dark region that is called a *black hole*. However, matter falling into a black hole must release some of its energy before it goes "inside." Thus there can be a bright region near the black hole. Further, the mass of the black hole still produces a gravitational effect. Radiation from infalling material and gravitational effects on the motions of nearby bodies can reveal the presence of a black hole and can give a measure of the mass that the black hole contains. In this way, black holes have been found with masses from a few times to a billion times the mass of the Sun. It has been shown that the center of the Milky Way contains a black hole with a mass of about 4 million solar masses.[1]

[1]A.M. Ghez, S. Salim, S.D. Hornstein, A. Tanner, M. Morris, E.E. Becklin, and G. Duchene, "Stellar Orbits Around the Galactic Center Black Hole," *Astrophysical Journal*, 620: 744-757 (2005).

with nearby galaxies, and the disruptive forces of these events trigger episodes of star formation. Meanwhile, the central black hole grows episodically by accreting material from the inner parts of the galaxy. The accretion disk that forms during such periods can sometimes produce more radiant energy than all the billions of stars in the galaxy combined—the black hole and disk in this condition is called an active galactic nucleus, or AGN.

An early result from radio astronomy was the realization that most of the bright sources of radio radiation lie outside our own Galaxy, the Milky Way, and have high redshifts, so they must be at "cosmological" distances. These objects lie in the nuclei of galaxies and are created as material swirls into giant black holes at the centers of the galaxies. Much of the radiation is emitted anisotropically in two narrow jets along the rotation axis of the black hole (Figure 3.4). The brightest objects—quasars—are those in which the jets are pointed almost directly toward Earth. The discovery and study of these powerful "radio galaxies" in the 1960s provided the first evidence for the existence of supermassive black holes—evidence that was based on the energy conversion required. A major discovery from radio astronomy, made by the technique of very long baseline interferometry (VLBI), was that the jets are flowing at relativistic speeds—close to the speed of light—and that the radiation is beamed by the effects of special relativity.

The best-studied supermassive black hole is the one in the center of the Milky Way; it has a mass of about 4 million times the mass of the Sun. Attention was first drawn to it as an important astronomical object in 1974, when radio emission from its envelope was seen. This radiation comes from relativistically excited gas that is

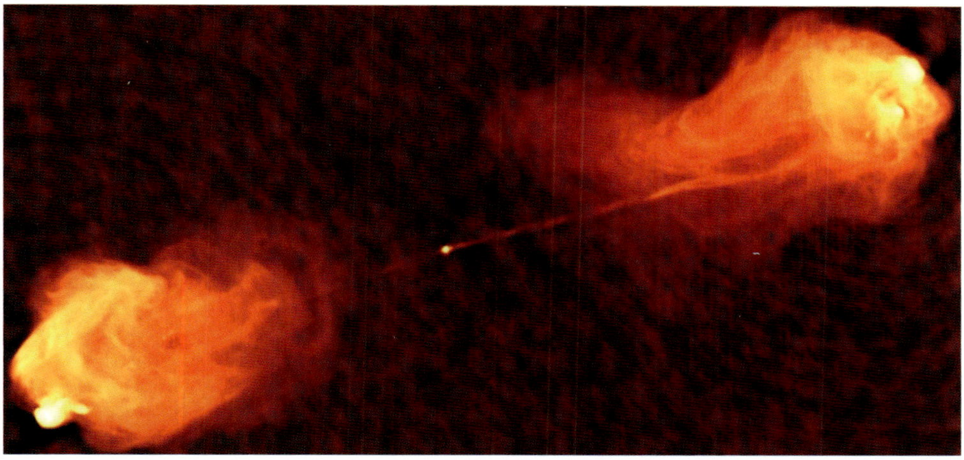

FIGURE 3.4 The remarkable nucleus, jets, and outer lobes of the radio galaxy Cygnus A. The nucleus contains a massive black hole that is accreting gas and dust, and some of the gravitational energy that is released is channeled into opposing jets. The jets contain a flow of relativistic plasma that, when stopped by the extragalactic material far outside the galaxy, generates the huge lobes. This image was made with the Very Large Array at a frequency of 5 GHz and with an angular resolution of 0.5 arcsecond. Image courtesy of NRAO/AUI/NSF.

spiraling into the black hole. It cannot be seen with an optical telescope because the central region is so dusty, but the radio waves readily penetrate dust.

The rate at which the black hole at the center of the Milky Way is growing has been measured by radio techniques. It currently is in a quiescent period, undergoing low accretion. In more active galaxies, the central black holes are accreting mass thousands of times faster.

Spectral line emission from water vapor at 22 GHz has turned out to be an unexpectedly important probe of the environments of supermassive black holes in the nuclei of galaxies. Water vapor appears as a trace constituent in the accretion disks that surround these black holes, and it emits radiation by the maser (microwave amplification by stimulated emission of radiation) process. This causes the emitting condensations, called spots, to appear as spectacularly bright but very compact sources of radiation whose positions and velocities can be measured precisely with a continental-scale radio telescope, called a VLBI array. In a stunning series of measurements, the orbital motions in the disk of one such galaxy, NGC4258, have been traced in detail (Figure 3.5). From these observations the mass of the black hole can be determined from Kepler's laws of motion, and also the distance of the black hole from Earth can be determined by the comparison of the angular and linear velocities of the maser spots. The measurement of distance by this direct trigono-

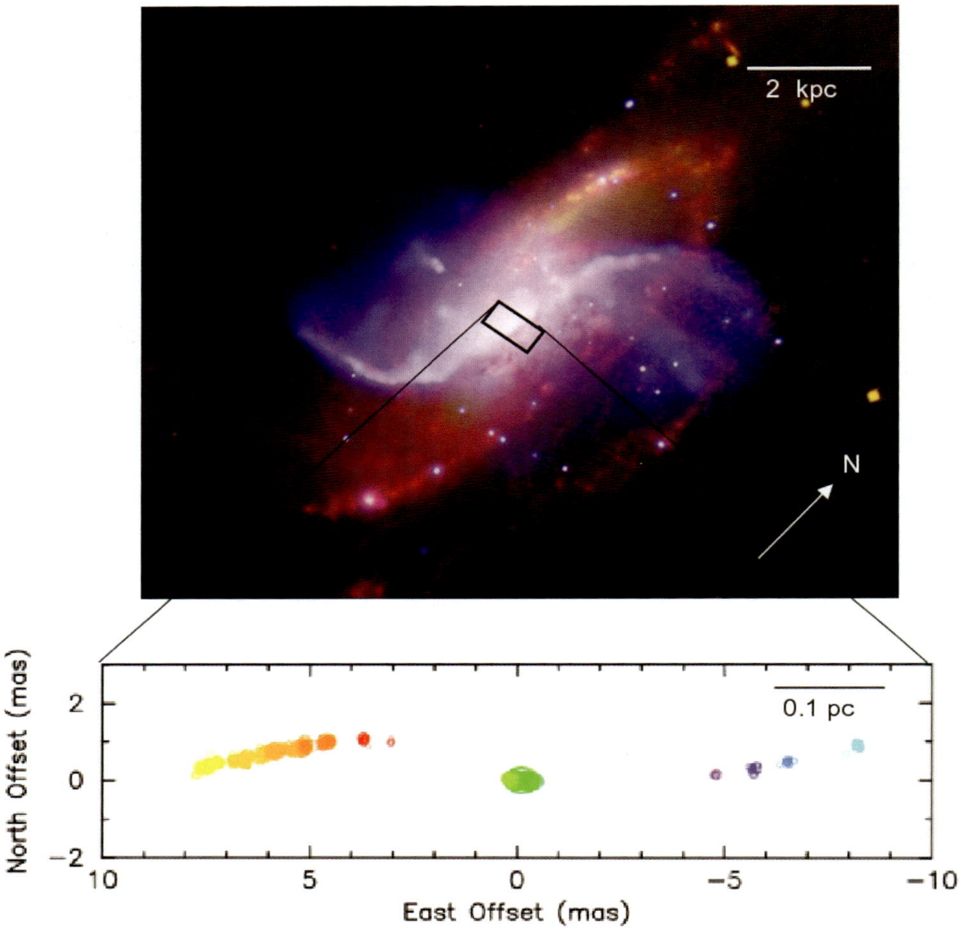

FIGURE 3.5 The galaxy NGC4258, shown in the top panel, is a relatively normal looking spiral galaxy, lying about 23,500 light-years from Earth. However, observations of the water line at 22 GHz show bright maser emission, as seen in the lower plot, whose scale is enlarged by a factor of 10,000 with respect to the upper plot. Each "spot" in the lower portion represents a separate maser whose velocity, derived from the Doppler shift, is color-coded: red = –500 km/s; blue = 1,500 km/s. The thin curved distribution of masers with the observed velocity distribution traces a thin disk of material in orbit around an unseen black hole with a mass of about 40 millions times that of our Sun. (Note: 1 pc ≈ 3.3 light-years.) Adapted from T. Yang, B. Li, A.S. Wilson, and C.S. Reynolds, "Spatially Resolved X-Ray Spectra of NGC4258," *Astrophysical Journal,* 660: 1106 (2007); and A.L. Argon, L.J. Greenhill, M.J. Reid, J.M. Moran, and E.M.L. Humphreys, "Towards a New Geometric Distance to the Active Galaxy NGC4258: I. VLBI Monitoring of Water Maser Emission," *Astrophysical Journal*, 659: 1040 (2007).

metric technique has important implications for establishing the "cosmic distance scale," that is, calibrating the relation between redshift and distance.[5]

As interesting as the active black holes themselves are the kinds of galaxies that give rise to such activity. Studies of the "host galaxies" in which active black holes reside have been made over the past few decades. Radio telescopes have been and will continue to be a major contributor to such studies, through their ability to detect star-forming gas and feedback from supernovae in host galaxies and other objects (see the subsection on "Galaxies," below). Both ALMA and EVLA, with their high resolution and sensitivity, will push the studies of star formation in host galaxies closer to the nuclear region in which the active black hole resides, thus allowing for the interplay between the black hole and nuclear star formation to be assessed.

Galaxies

The study of star formation in our Galaxy and in others is one of the primary areas of science done at millimeter wavelengths (68-115 GHz). Stars form in giant molecular clouds composed primarily of diatomic hydrogen (H_2); however, H_2 is particularly difficult to detect because it has no permanent dipole moment. As a result, astronomers use carbon monoxide (CO) as a proxy for H_2. CO is collisionally excited by H_2, and the resultant emission from CO is observable at radio wavelengths. Hence the properties of star-forming gas are commonly measured with radio telescopes. An example of CO emission from a nearby galaxy is shown in Figure 3.6.

CO emission has been detected in many varieties of galaxies, including some with redshifts up to 6.4, so that the photons now observed were emitted when the universe was only a few percent of its present age. With radio telescopes, it is thus possible to study the properties of star formation in normal galaxies such as our own, in exotic galaxies with vigorous star formation accompanied by accretion onto black holes (e.g., radio galaxies and quasar host), and in distant galaxies likely undergoing their first burst of star formation. An important fact about these observations is that, owing to the motions of nearby galaxies and the redshifts of more distant ones, CO emission is rarely observed at or even near the rest frequency. The 115 GHz line is observed in "local" galaxies (redshift <0.3) down to frequencies of 88 GHz. Observations at high redshift ($z > 2$) are becoming routine; this requires either looking at higher-level transitions of CO redshifted into the 3 mm (68-115 GHz) window or observing the ground-state (115 GHz) transition at much lower frequencies (22-50 GHz).

---

[5]J.R. Herrnstein, J.M. Moran, L.J. Greenhill, P.J. Diamond, M. Inoue, N. Nakai, M. Miyoshi, C. Henkel, and A. Riess, "A Geometric Distance to the Galaxy NGC4258 from Orbital Motions in a Nuclear Gas Disk," *Nature*, 400: 539 (1999).

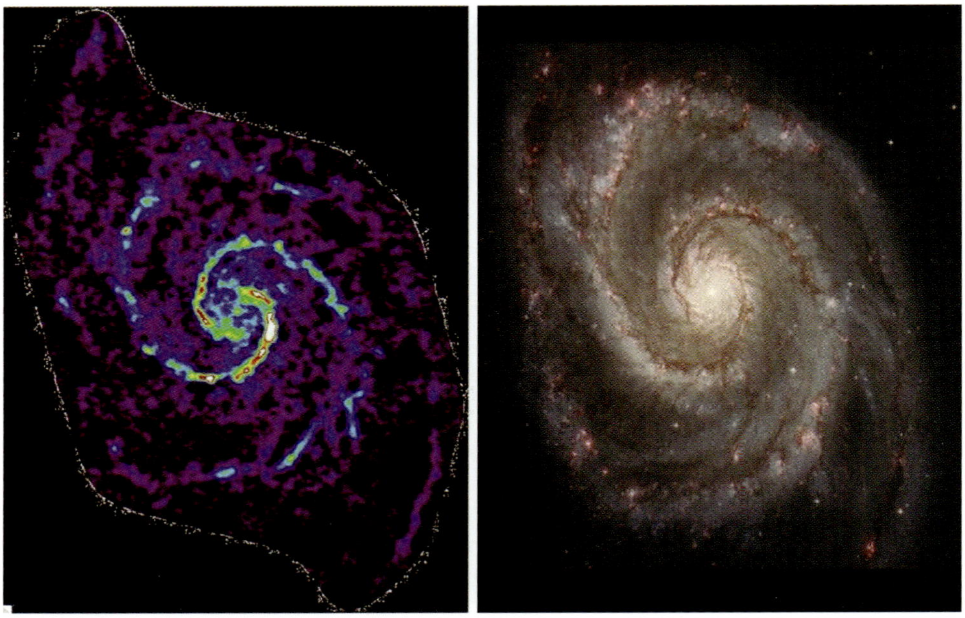

FIGURE 3.6 (Left) An aperture synthesis map at 115 GHz of carbon monoxide (CO) spectral line emission from the Spiral Galaxy Messier 51 (the "Whirlpool" Galaxy). The CO, which is tracing star-forming molecular gas, is observed to follow the spiral arms shown in the Hubble Space Telescope optical image of the galaxy (right). The image is approximately 40,000 light-years across. The CO image was made by combining 200 hours of observations at the Combined Array for Research in Millimeter-wave Astronomy with 40 hours of observations at the Nobeyama Radio Telescope in Japan. Image courtesy of Space Telescope Science Institute.

Solar Physics and Space Weather

The nearby Sun is the only star that there is a chance of studying in detail. Knowledge of the Sun illuminates the understanding of other stars and generally helps place the Sun and its attendant suite of planets into the context of stellar physics and the evolution of stars and planets. In addition, the Sun's atmosphere is a remarkably active, even violent region, and it regularly impacts Earth with disturbances that can have technical and economic consequences. There currently is a proposal to build a powerful new instrument, the Frequency Agile Solar Radiotelescope (FASR), that would greatly increase the capability of measuring the solar atmosphere over a wide frequency range, at high time and angular resolution.

The Sun's atmosphere emits strongly at all radio frequencies by a variety of emission mechanisms, allowing observers to probe the physical processes that are active on the Sun. Flares on the surface are explosions connected with the disappearance or reconnection of magnetic fields (Figure 3.7). Strong bursts of

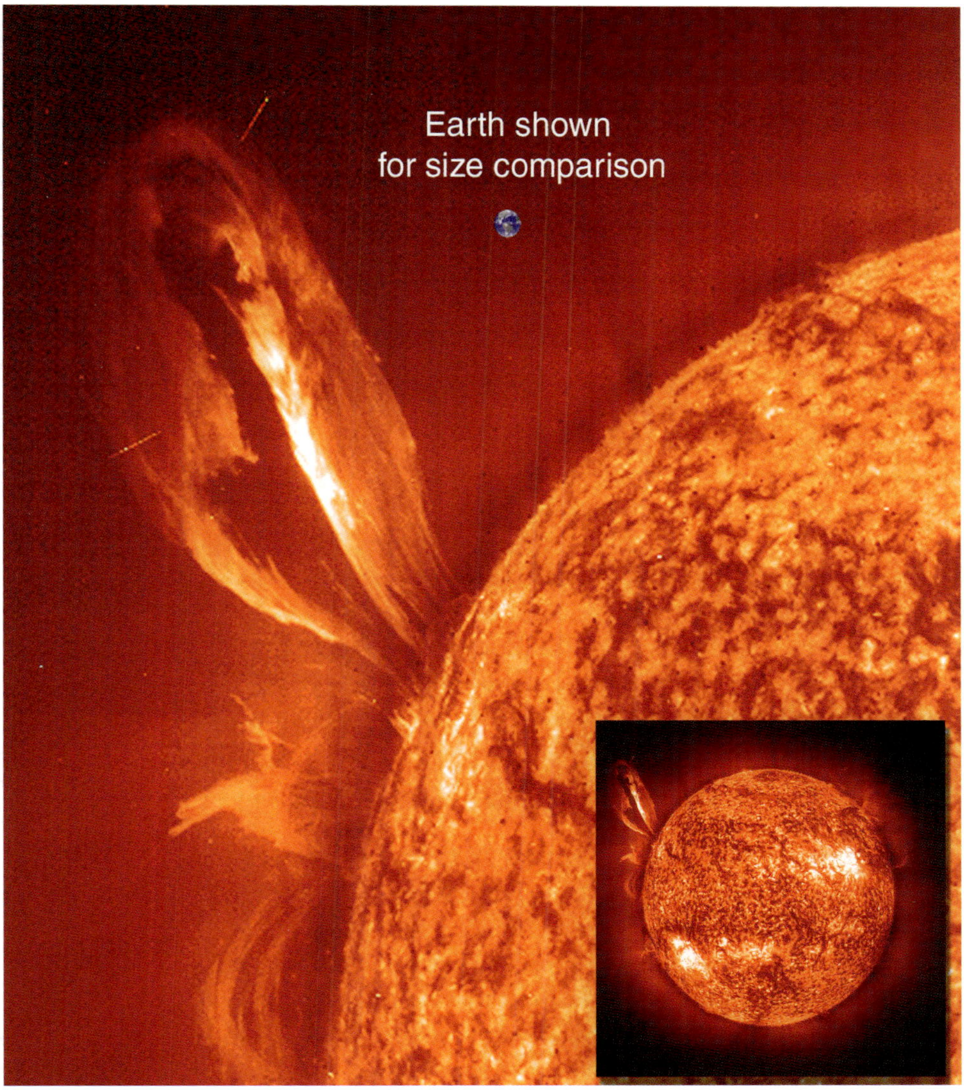

FIGURE 3.7 A large eruptive prominence above a solar flare, seen in the ultraviolet light of ionized helium, with the Solar and Heliospheric Observatory (SOHO) satellite, July 24, 1999. For size comparison, Earth is shown as the small blue circle. The flare started with an eruption of twisted magnetic field through the surface. The magnetic field loop is rising rapidly through the corona and will separate from the Sun to form a coronal mass ejection (CME). This particular CME did not hit Earth, however, as it started in a direction perpendicular to Earth. The inset shows other active regions on the face of the Sun. Image courtesy of SOHO (ESA and NASA).

radio noise are often associated with a flare, and indeed on December 6, 2006, the radio bursts were so intense that for 10 minutes they disrupted the Global Positioning System (GPS) reception on essentially the entire sunlit side of Earth.[6] An associated phenomenon, coronal mass ejections (CMEs), involves the eruption of mass and magnetic flux from the Sun into interplanetary space. These can strongly disturb the near-Earth environment. There is general agreement that flares and CMEs are magnetic phenomena, but the details are unclear.

Space weather refers to the highly variable condition of the plasma that surrounds Earth and extends from the Sun throughout the solar system. The solar wind is a continuous stream of plasma that blows out from the Sun, and it controls the shape of the outer regions of Earth's magnetic field. Flares sometimes produce energetic particles that propagate to Earth in a matter of minutes.[7] Similarly, a CME can also produce energetic particles. These energetic particles can be a danger to the personnel and equipment in space vehicles. The CME itself takes 1 or 2 days to travel to 1 AU (astronomical unit; the Sun-Earth distance). If it hits Earth, it can cause serious communication disturbances and adversely affect satellites and long-distance high-voltage transmission lines.[8] Because of these disruptive consequences, it is important that as much as possible be learned about flares and CMEs and that the capability of predicting them be developed. Much of this study must be done at radio wavelengths, although the radio information is supplemented with data from other wave bands—for example, x-rays measured from a satellite.

Serendipity and the Transient Universe

Throughout astronomy, in optical and other bands as well as radio, we are entering an era of intense surveillance, variously called transient source astronomy, time domain astronomy, and, more broadly, a new frontier in high energy astrophysics. The objective is to capture transient phenomena, which currently are enjoying wide attention. Long-known transient phenomena include novae and supernovae, pulses from pulsars, and motions in quasars and galactic nuclei, in addition to solar system phenomena such as eclipses and solar flares.

New radio telescopes, such as the ATA, will be used in a repetitive survey mode to search for transient and variable events. These phenomena are of broad significance and will contribute to the understanding of the life and death cycle of stars; the nature of exotic compact objects such as neutron stars, white dwarfs,

---

[6]R. Cowen, "Big Broadcast," *Science News*, 171(23): 360 (June 9, 2007).

[7]See http://www.nasa.gov/home/hqnews/2005/may/HQ_05132_solar_fireworks.html, accessed May 21, 2008; R.A. Mewaldt, "Solar Energetic Particle Composition, Energy Spectra and Space Weather," *Space Science Reviews*, 124: 303-316 (2006).

[8]See http://ds9.ssl.berkeley.edu/solarweek/WEDNESDAY/spaceweather.html; accessed May 21, 2008.

and black holes; and the physics of magnetized, relativistic plasmas. Transient and variable phenomena are typically broadband, occurring at all radio frequencies and on timescales from nanoseconds to years. Multifrequency, repetitive observations are necessary to characterize the physics of these targets.

Among the projected science targets are phenomena connected with explosions of massive stars, which might produce short bursts of powerful radio emission. The discovery of such events could confirm the fundamental picture of what is called the gamma-ray burst phenomenon and provide an independent method for the discovery of distant star-forming galaxies. Radio studies of the propagation effects that these waves encounter will probe the very tenuous intergalactic medium that constitutes a significant fraction of the baryonic content of the universe.

Neutron stars that emit sporadic pulses have been found very recently, and estimates have been made that such stars are abundant in the Galaxy. These objects are likely providing new insights into physical conditions in neutron star magnetospheres.

Magnetic activity on the surfaces of stellar and compact objects belongs to a continuum of activity that includes solar flares. A comprehensive census of this activity and the detection of true solar-like events on other stars will provide important insights into the physics of solar flares as well as identifying conditions suitable for life on extrasolar planets.

Repetitive surveying at high time and frequency resolution is a new regime in astronomical *phase space*—that is, the parameter space representing all possible observations. In the past, opening such a new regime generally has led to dramatic new, often unexpected, discoveries. Scientists cannot predict what will be found, but on the basis of past experience, they do expect to see new phenomena. The ability to distinguish between transients of cosmic origin and sporadic radio frequency interference will be a challenging enterprise.

Summary

Radio astronomy has provided astronomers a unique way to observe and analyze cosmological objects of interest, from Earth's Sun, to galaxies, to the very beginning of the universe itself. The field has thus been responsible for some of the most important astronomical findings to date. As capabilities increase and new observatories come online, radio astronomy is poised to allow scientists to understand the universe in unprecedented ways.

Finding: Radio astronomy has great potential for further fundamental discoveries, including the origins and evolution of the universe, the nature of matter, and life in other solar systems, which will have an enormous impact on our understanding of fundamental physics and the place of humanity in the universe.

3.2 RADIO OBSERVATORIES AND RADIO TELESCOPES

Radio observatories contain a diverse group of scientific instruments carefully designed and built to observe with the highest sensitivity selected aspects of the radio emission from the many varieties of objects in the universe. No single instrument, observatory, or even observing technique can encompass the broad frequency range (tens of MHz to hundreds of GHz), the wide range of angular scales (tens of micro-arcseconds to degrees), and the broad range of temporal variations (nanoseconds to many years) that are seen in the emission. Hence, radio observatories have a variety of telescopes and instruments with unique technical capabilities. Box 3.4 describes several current and future radio astronomy observatories (three are in operation and one is under construction).

At the lowest frequencies, telescopes consist of dipoles (simple lengths of wire or metal) or arrays of dipoles linked together. These structures are simple, cheap, and efficient. Above 100 MHz, telescopes take on the classic parabolic shape but can be surfaced with wire mesh. The mesh saves money and weight in the telescope; the radiation is efficiently collected because the wavelength is much larger than the holes in the mesh. At about 1 GHz and higher, the telescopes need highly precise solid surfaces and stable guiding structures. Two telescope systems currently being designed—FASR and SKA—are examples of the design being matched to the frequency. The plan for both FASR and SKA is to use dipole arrays to cover frequencies below 300 MHz, low-precision parabolic reflectors to cover from 300 MHz to 3 GHz, and high-precision parabolic reflectors from 3 to 30 GHz.

Techniques are different at the highest frequencies, 30-1,000 GHz, where quasi-optical techniques are often used: that is, signals are directed through mirrors to the detectors rather than through waveguides. At the extreme high frequency end, the required surface accuracy of reflectors is about 15 microns, one-fifth the diameter of a human hair.

Angular resolution, the ability to image fine structure, is a second factor driving telescope design. The resolution is determined by the ratio of the wavelength of observation to the diameter of the telescope. Depending on the science objectives, it may be desirable to have arcminute or even sub-milli-arcsecond resolution; however, getting very high resolution by building an extremely large dish is impractical. For example, the Arecibo telescope, at 305 meters in diameter, is the largest dish-type telescope in the world (see Figure 3.4.3 in Box 3.4). Its highest operating frequency is 10 GHz, where it has a resolution of about 30 arcseconds. Getting more resolution at this frequency by building a larger dish would be much more expensive than building a linked array of smaller telescopes in which the resolution is controlled by the overall size of the array. The VLA in New Mexico has 27 telescopes that can form a baseline up to 35 kilometers, giving the VLA a resolution of 0.3 arcsecond at 10 GHz (see Figure 3.4.4 in Box 3.4). The Very Long

BOX 3.4
Radio Astronomy Observatories

FIGURE 3.4.1 The Robert C. Byrd Green Bank Telescope (GBT) at the National Radio Astronomy Observatory in Green Bank, West Virginia. With a diameter of 100 meters, the GBT is the world s largest fully steerable telescope. It operates from 300 MHz to 90 GHz and is predominantly used for radio spectroscopy and for studies of pulsars. It has an offset feed support system to eliminate radio shadows on the dish, which can be troublesome when sensitive measurements are being made. The GBT and Green Bank, West Virginia, are in the National Radio Quiet Zone (NRQZ); see §3.6 in this report. Image courtesy of NRAO/AUI/NSF.

continued

BOX 3.4 Continued

FIGURE 3.4.2 Artist's conception of the Atacama Large Millimeter Array (ALMA) now being built in the Atacama desert in northern Chile. When completed, ALMA will have up to 80 antennas, operating from 30 GHz to 960 GHz. ALMA is at an altitude of 5,000 meters, where the atmospheric water vapor is low enough that these high frequencies are useable. This project is a collaboration of the United States, Canada, the European Southern Observatory, and Japan. Note in this figure that the individual telescopes are not identical: the one on the left has a European design, and that on the right has an American design. Image courtesy of NRAO/AUI and Computer Graphics by ESO.

FIGURE 3.4.3 The 305 meter Arecibo telescope at Arecibo Observatory in Puerto Rico. Built in 1963 and operated by the National Astronomy and Ionosphere Center (NAIC), the Arecibo telescope still has the largest collecting area of any radio telescope in the world. It has undergone several major renovations, including the installation of a complex secondary feed system (inside the white enclosure) that corrects for the fact that the primary reflector is a section of a sphere, not a paraboloid. It operates from 300 MHz to 10 GHz, with continuous frequency coverage above 1.1 GHz. The large foreground building is the Angel Ramos Visitor Center, which receives more than 100,000 visitors per year. Image courtesy of the NAIC-Arecibo Observatory, a facility of the NSF.

continued

BOX 3.4 Continued

FIGURE 3.4.4 The Very Large Array (VLA) near Socorro, New Mexico, consists of 27 antennas, each 25 meters in diameter, connected as an interferometer to produce radio images at frequencies from 70 MHz to 43 GHz. The antennas are in a "Y" pattern and can be repositioned to different configurations, with a maximum baseline of 35 kilometers, to produce images of various angular resolutions. It currently is being upgraded to have more sensitivity and better image quality. In its new state, the Expanded Very Large Array, it will have continuous frequency coverage from 1 to 50 GHz. Image courtesy of NRAO/AUI/NSF.

Baseline Array (VLBA), with maximum baselines of 6,000 to 8,000 km, has an angular resolution of roughly 0.1 milli-arcsecond at 43 GHz.

In practice, angular resolution, operating frequency, and total collecting area are considered jointly in optimizing the design of a telescope. Different possible solutions to the structure usually exist, and one is chosen according to the primary science goals for the observatory. Increasing the angular resolution can assist in reducing the potential for interference. However, the design process is quite complex if the angular resolution is to be maximized while the sidelobes that capture interfering signals are to be minimized.

Another key factor in optimizing the capability of an observatory is its location, and the broad spread of the radio spectrum results in a number of factors that can be important. At frequencies below 30 GHz, radio frequency interference (RFI) is an important cause of noise and signal degradation. The National Radio Quiet Zone in West Virginia, where the GBT is located, is important because there is a legal and effective means of minimizing RFI there (see Figure 3.4.1 in Box 3.4). At high frequencies, water vapor in the atmosphere is an important source of noise and attenuation. ALMA and other telescopes are being built at an elevation of 5,000 m in the Atacama Desert in Chile to optimize their performance up to 1000 GHz (see Figure 3.4.2 in Box 3.4).

A highly sensitive receiver, or radiometer, is coupled to the radio telescope. At frequencies below about 50 GHz this is a low-noise amplifier, usually containing a cooled transistor. Transistor technology continues to improve, however, and the upper frequency limit for transistors' use has been rising steadily. Above 50 GHz, more complicated devices are used, including superconductor-insulator-superconductor (SIS) junctions. In addition, above 100 GHz, bolometers are commonly used, especially for broadband continuum measurements. Focal plane arrays, both of bolometers and coherent devices, are coming into regular use. An array of detectors is essentially a radio camera, with from a few pixels to hundreds of pixels—far fewer than a modern digital camera has but still, such a radio camera will operate 100 times faster than a conventional system with a single point feed. An interferometer system is automatically such an array, and its ability to form an image with many pixels is limited only by its computing power (and the primary beam of the antenna elements).

The signal that comes from the radiometer can be used in various ways. It can be directly detected as a broadband signal to maximize sensitivity to thermal or synchrotron emission. It can be closely sampled in time to search for pulses from neutron stars or used to construct a spectrum for the study of molecular or atomic spectral lines. The astronomy signals are almost always a very small fraction of the internal noise in the receiver and can only be measured by using a long integration time, sometimes of many hours (see §3.4).

Table 3.1 highlights selected major radio observatories currently operating,

TABLE 3.1 Selected Major U.S. Radio Observatories Around the World: Operating, Under Construction, Being Planned

Observatory	Location	Frequency (GHz)	Collecting Area[a] (m²)
Selected Operating Facilities			
Allen Telescope Array (42 dishes)	Hat Creek, California	0.5-11.2	1,230
Arecibo Observatory	Arecibo, Puerto Rico	0.3-10	73,000
Arizona Radio Observatory	Tucson, Arizona	68-500	78 and 113[b]
Atacama Cosmology Telescope	Chile	150-270	28
Caltech Submillimeter Observatory	Mauna Kea, Hawaii	200-950	85
Combined Array for Research in Millimeter-Wave Astronomy	Owens Valley, California	70-260	770
Green Bank Telescope	Green Bank, West Virginia	0.3-100	7,850
Large Millimeter Telescope	Mexico	85-275	1,960
South Pole Telescope	South Pole	95-275	78
Submillimeter Array	Mauna Kea, Hawaii	180-900	226
Very Large Array	Socorro, New Mexico	0.07-50	13,250
Very Long Baseline Array	10 sites in United States	0.3-90	4,900
Selected Facilities Under Construction			
Allen Telescope Array (350 dishes)	Hat Creek, California	0.5-11.2	10,220
Atacama Large Millimeter Array	Chile	30-960	6,000
Long Wavelength Array 1+	New Mexico	0.015-0.09	20,000 @ 15 MHz
Murchison Widefield Array	Murchison, Australia	0.08-0.3	8,000
Selected Facilities in Planning			
Cornell Caltech Atacama Telescope	Chile	200-900	490
Square Kilometer Array	To be determined	To be determined	1,000,000

[a]The collecting area listed in column 4 is the geometric area of the aperture for the dish-type telescopes.
[b]Two telescopes of 10 and 12 m diameter.

facilities under construction, and facilities being planned within the U.S. community. The operating observatories represent an investment of roughly $1 billion. Some of the newest observatories will be built in collaboration with other countries, a trend that will increase in the future. ALMA, a $1 billion observatory under construction in northern Chile, is a collaboration among institutions in North America, Europe, East Asia, and Chile. The Square Kilometer Array, a project currently being designed and prototype tested, is a world collaboration that is also expected to cost more than $1 billion to build.

Note that a third of the facilities listed in Table 3.1 are not located in the United States, although they are supported and operated in part or completely by U.S. public and private institutions. The Atacama Cosmology Telescope (ACT), Large Millimeter Telescope (LMT), Submillimeter Array (SMA), South Pole Telescope

(SPT), ALMA, and Cornell Caltech Atacama Telescope (CCAT) are at high altitude to minimize the difficulties produced by atmospheric water vapor. The Murchison Widefield Array (MWA) is in Western Australia, where currently the RFI is exceptionally low. The Australian and South African governments have established a level of protection against RFI for the SKA in the event that it is built in their respective countries. The Chilean government has done this for ALMA, which is now under construction.

Finding: Scientific advances have required increasing measurement precision by passive radio and microwave facilities in order to obtain more accurate and thus more useful data sets. This need for precision will continue to increase.

Finding: Large investments have been made in satellite sensors and sensor networks and in major radio observatories. New facilities costing billions of dollars are under construction or are being designed.

3.3 SPECTRUM REQUIREMENTS AND USE

The spectral windows used to observe cosmic objects of interest are determined by the physics of the objects and the atmosphere through which the incoming radiation must pass. Using current spectrum allocations as well as many windows of opportunity, radio astronomers are able to learn fascinating information about the cosmos in which we live.

Continuum and Line Observations

Most radio astronomy observations fall into one of two categories: continuum observations and line observations. With respect to continuum observations, continuous spectrum from a radio source covers a wide frequency range, often a factor of about 1,000 in frequency, and the intensity commonly changes slowly with frequency. The spectrum often shows a maximum in some band, but some sources show a steady change, either increasing or decreasing, with frequency, over the entire available radio range. The sensitivity of continuum observations is proportional to the square root of the receiver's bandwidth (see §3.4), so often the bandwidth is made as wide as is practical, limited by the technology of the receiver and by external interference. As an example of technology-limited bandwidth, consider very long baseline interferometry. In this case, signals from multiple, separated antennas are recorded for later processing. In 1967 the first VLBI system used a 330 kHz band because that was all that was available with computer tape drives. The bandwidth steadily increased as better recording systems became available, and now recordings at more than 1 GHz are made, on hard disks, at frequencies above 10 GHz. The objective has been increased sensitivity. Increased sensitivity translates into a larger portion of the universe that can be studied, because of the squared (r^2) distance effect. Improving the sensitivity will be a strong driver for radio astronomy equipment for a long time to come.

Modern continuum observations cannot be restricted to the bands allocated to the Radio Astronomy Service (RAS); wider bands are needed for sensitivity. Meeting this need has another effect, however; it increases the exposure to RFI. This problem will worsen with time as transmissions increase and the sensitivity of radio systems continues to be improved.

Line observations refer to the radiation in spectral lines from quantum transitions of atoms or molecules. Different transitions give different linewidths, but they are well under 1 percent of the frequency. Hence specialized, narrowband receivers are used. The observations, however, must be made at the transition frequency regardless of what RFI is present there. The most famous spectral line, arguably the most important one for radio astronomy, is the atomic hydrogen line at 1420 MHz. This line is protected, with the 1400-1427 MHz band allocated to the RAS on an exclusive primary basis. Even so, RFI has been seen in this band. At high frequencies, especially above 100 GHz, broad bandwidths are often used in this application to encompass many spectral lines simultaneously.

While broad bandwidths are often used at millimeter-wave frequencies to encompass many spectral lines simultaneously, recently there has also been renewed interest in making wide-bandwidth spectral scans at lower frequencies.[9] Most of the frequency spectrum observed in these surveys has no protection against RFI. Of course, observations of spectral lines in external galaxies rarely fall in protected bands due to the redshift of the target, even when the rest frequency of the line is protected.

Pulsar observations are in a different category because they emit short pulses that can only be seen with a short integration time. These short pulses can be co-added with appropriate time shifts, like radar pulses, to enhance sensitivity. In addition, the pulses drift in frequency owing to intervening dispersive plasma. Multichannel observations are required to limit dispersive smearing, and voltage-based signal processing is implemented to remove dispersion effects.

[9]For example, the Prebiotic Interstellar Molecule Survey, a large-scale search for new organic molecules from the Sagittarius B2 region using the GBT between 300 MHz and 50 GHz; also, spectral scans with almost complete coverage from 1 to 10 GHz have been made on both galactic and extragalactic targets from Arecibo (C.J. Salter, T. Ghosh, B. Catinella, M. Lebron, M.S. Lerner, R. Minchin, and E. Momjian, "The Arecibo Arp 220 Spectral Census I: Discovery of the Pre-Biotic Molecule Methanimine and New Cm-wavelength Transitions of Other Molecules," *Astronomical Journal*, 136(1): 389-399 (2008)).

Atmospheric Windows and Absorption Features

The allocation of spectral bands for radio astronomy is based partly on the available atmospheric transmission windows, as shown in Figure 1.3 in Chapter 1. Ground-based telescopes can observe only in bands where the atmosphere does not absorb the radiation. Starting at the ionospheric cutoff near 15 MHz and extending to about 50 GHz is a relatively clear band. Above 50 GHz, radio windows occur approximately at 65-115 GHz, 125-180 GHz, and 200-300 GHz. At still higher frequencies the windows are less distinct, but they do exist at 330-370 GHz, 460-500 GHz, 600-700 GHz, and 800-900 GHz, as well as in other, narrower windows.

Current Radio Astronomy Service Allocations

Figure 3.8 shows the frequency bands that currently are allocated to the RAS. The U.S. and international spectrum allocation table and footnotes are available in the National Telecommunications and Information Administration's (NTIA's) *Manual of Regulations and Procedures for Federal Radio Frequency Management (Redbook)*[10] and in the Federal Communications Commission's (FCC's) *Frequency Allocation Table.*[11]

RAS has a narrow band approximately every octave across the radio spectrum, which allows the investigation of both the broadband and the spectral line emissions of celestial sources. Band allocations start at 13.4 MHz and extend to 275 GHz, as shown in Figure 3.8.

Spectrum Use

The following subsections discuss the spectrum divided into five broad ranges: below 100 MHz, 100-1420 MHz, 1.4-30 GHz, 30-275 GHz, and 275-3000 GHz. A brief description of some of the current major scientific investigations in each range is presented, and new uses expected in the coming decade are mentioned.

Below 100 MHz

In general, continuum sources will come under study in the spectrum range below 100 MHz. These sources include the Sun and Jupiter, as well as other stars and, possibly, Jupiter-like planets around other stars. Many extragalactic

---

[10] Available at http://www.ntia.doc.gov/osmhome/redbook/redbook.html; accessed January 13, 2010.

[11] Available at http://www.fcc.gov/oet/spectrum/table/; accessed January 13, 2010.

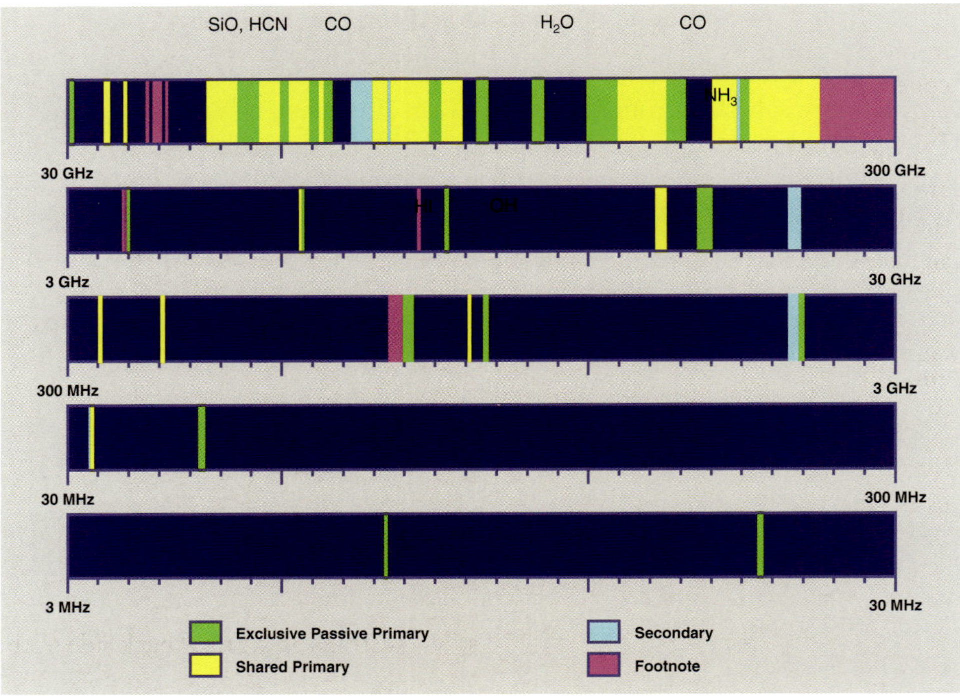

Radio Astronomy Frequency Allocations in the United States

FIGURE 3.8 Spectrum allocations to the Radio Astronomy Service (RAS) covering the range 3 MHz to 300 GHz. Exclusive primary bands are in green, and shared primary bands are in yellow. The spectral region above 75 GHz is widely used by the passive services but little used by the active services for lack of suitable technology. That will change in the future. Some of the bands at lower frequencies are more threatened by radio frequency interference than others. For example, the band 23.6-24.0 GHz, along with several bands between 22.0 and 23.6 GHz (not shown), are used by the RAS and the Earth Exploration-Satellite Service because they contain the important water vapor and ammonia spectral lines. Automotive collision-avoidance radar, which will come into wide use in the next few years, will be in the band 22-27 GHz. The potential for interference is high. Image courtesy of Andrew Clegg, National Science Foundation.

sources have steep synchrotron spectra such that they are most powerful at low frequencies.

A new radio telescope, the Long Wavelength Array (LWA), is now under construction in New Mexico. LWA observations will complement those at higher frequencies. The observatory is planned to consist ultimately of 53 stations spread over 400 km; each station will contain 256 broadband dipoles operating from 10 to 88 MHz. The angular resolution of the LWA will be a few seconds of arc, and the instantaneous field of view will be a few degrees. The high-resolution, low-frequency possibilities that these capabilities open up represent a new regime in radio astronomy. Some plasma regions, including pulsar atmospheres, radiate coherently at low frequencies, and discovering these or other transient objects may give the most significant results from the early use of this instrument.

Interference is particularly severe at frequencies below 100 MHz, where there are many commercial and government services, both fixed and mobile. Although the beam-forming nature of the system will automatically reject some interfering signals, there remains the strong potential for RFI. The LWA design will implement a variety of RFI mitigation procedures.

100-1420 MHz

Studies of hydrogen, the most abundant element in the universe, are particularly important in the 100-1420 MHz range. Pioneering efforts are underway using the 1420 MHz spectral line of hydrogen, to detect material that is heated by the first generation of stars in the early universe. That radiation now must be observed at much lower frequencies, owing to the large redshift. The radiation will be spread over a very broad band, and the signals will be particularly weak. Months of integration at remote sites, such as Western Australia and/or the back side of the Moon, will be required for reliable detections. A major instrument, the MWA, is now under construction in Western Australia for this purpose.

The hydrogen line at 1420 MHz is used to study the motions and dynamics in the Milky Way and in external galaxies out to great distances. "Dark galaxies" with much hydrogen but few stars are also expected to exist. They will form a new frontier for observation, at frequencies from 1420 MHz down to about 300 MHz.

The heavy isotope of hydrogen, deuterium, has an analog of the 1420 MHz line at 327 MHz. This line, first detected only a few years ago, will be an important subject of study in the coming decade. It will provide information related to the origin of the universe and the cosmological synthesis of the elements.

One of the most interesting and significant discoveries in radio astronomy was the detection of pulsars. Their huge magnetic, electric, and gravitational fields, impossible to reproduce in laboratories on Earth, allow observations of matter and radiation under extreme conditions. Pulsars generally emit most strongly at frequencies in the range from 50 to 600 MHz, but they are often observed up to a few gigahertz and, for a few objects, to 100 GHz.

1.4-30 GHz

The study of the nuclei of galaxies, including that of the Milky Way, is an important and fundamental topic in astronomy; it is done to a large extent between 1.4 and 30 GHz. Problems that can be studied in this range include the properties of massive black holes, explosive activities and the production of intense double radio sources from galactic nuclei, the collimation and acceleration of relativistic jets of plasma, the influence of galactic nuclei on the morphological structure of galaxies, and the formation of galaxies and quasars.

The study of hydroxyl (OH) with primary bands at 1.6-1.7 GHz and also at 4.7 GHz, 6.0 GHz, and other frequencies is of interest for investigating phenomena associated with the formation of protostars and the initial stages of star formation. OH is often seen in the form of masers in the atmospheres around stars. Exceedingly strong emission from OH "megamasers" is seen in some galaxies. It can be a million or more times stronger than the emission from masers in the Milky Way and so can be seen to great distances. These observations give information on magnetic fields in other galaxies and on their evolution over cosmic time.

30-275 GHz

The spectral region above 30 GHz is crucial for the identification and study of interstellar molecules. Some astronomers and biologists think that interstellar chemistry may have supplied Earth with prebiotic compounds essential for terrestrial life. Consequently, establishing the inventory of molecules in interstellar gas is central to astrobiology and astrochemistry. In addition, molecules in the 30-275 GHz frequency range provide essential diagnostics for star formation.

The band from 65 to 115 GHz has relatively little absorption from the atmosphere and is one of the best for both continuum and spectral observations. More than 100 molecules, as well as 25 different isotopic species, have been detected here. These include complex molecules such as CH_3 CH_2 OH and CH_3 OCH_3. Some molecules have several isotopic species in this range, so that isotopic abundance ratios can be studied. As an example, the basic molecule HCN has the isotopic species H_{12} C_{14} N, H_{14} CN, and H_{12} C_{15} N in the 86-92 GHz range, and all have been observed in the interstellar gas. The most important transitions in this frequency range, however, are generated by the CO molecule at 115 and 230 GHz. Emission from these lines is pervasive throughout our entire Galaxy and in other galaxies. Indeed, the bulk of the literature here is based on CO observations. CO's millimeter transitions are widely used to trace star-forming molecular gas, and this is crucial for assessing star formation in the Milky Way and other galaxies. As the sensitivity of telescopes continues to improve, studies of star formation using the weaker lines of HCN and HCO+ are also becoming important.

The band near 43 GHz is regularly used to study quasars and galactic nuclei with the VLBA. At this high frequency, extreme angular resolution is obtained, 0.1 milliarcsecond. This corresponds to a footprint on the Moon, as seen from Earth.

The frequency band 217-231 GHz provides a window near the peak of the CMB spectrum. Because of its low intensity and the strong variable contaminating emission from the atmosphere, accurate measurement of the CMB must be made in extreme environments, with high-altitude radio telescopes, at the South Pole, or with high-altitude aircraft, balloons, and spacecraft.

This region of the spectrum has become increasingly important in the past two decades, and the emphasis placed on this band will continue to increase as new telescopes and new instrumentation proliferate.

275-3000 GHz

Exploration of the electromagnetic spectrum between 275 and 3000 GHz has only begun in earnest in the past decade, as a consequence of the great strides made in the development of quantum heterodyne mixers and high-precision large-aperture antennas and the ability to make large arrays from them. Because the water vapor in the atmosphere is only partially transparent in selected portions of this band (see Figure 1.3), observations must be done at extraordinarily dry sites, most of which are at elevations greater than 4,000 m. The peak of the entire electromagnetic spectrum of the universe occurs in the middle of this band, at about 1 THz.

Extraordinary opportunities exist to study the universe in the early stages of its development, especially around redshifts of about 6-10 when the first stars reionized the universe at the end of the so-called Dark Ages. An important concept is that the intensity of thermal radiation from galaxies, which follows the Rayleigh-Jeans law, is proportional to the square of the frequency, so that galaxies' measured flux densities are essentially independent of distance, because the increasing redshift of the radiation due to the expansion of the universe exactly compensates for the inverse-square law loss suffered in propagation. The first deep images from the ALMA array (see Table 3.1), now under construction, are expected to be dominated by galaxies at great redshift that are not seen at all in the deep field images of the Hubble Space Telescope at optical wavelengths.

The 275-3000 GHz band will be very important in the field of astrochemistry, which seeks to understand how various molecules form and build up in complexity in regions of the interstellar medium where dense molecular clouds form and spawn new generations of stars. The importance of this band is due to the fact that the intrinsic strength of spectral lines from molecules and atoms increases as the fourth power of frequency. Hence the spectrum in this region is almost a "forest" of spectral lines in the direction of star-forming molecular clouds. Instruments such as ALMA will be able to image these regions with high angular resolution that will only be surpassed by infrared arrays in space, which are many decades from feasibility.

A critical astronomical problem of our age is the question of how planets form from the debris disks left over after a star forms from it host molecular cloud. The emission strength of the dust in such disks increases as the square of the frequency and is most readily imaged at the highest radio frequencies, which are afforded in the 275-3000 GHz band.

In addition, the radio source associated with the supermassive black hole in the center of the Milky Way has a peak in its emission spectrum at about 600 GHz. This source is obscured by plasma scattering at frequencies below 200 GHz, and it can only be studied directly at higher frequencies. The size of the source has recently been determined to be 37 micro-arcseconds from VLBI observations.[12] Observations at higher frequencies with larger VLBI arrays will provide images that show how light is bent in the strong gravity regime close to the event horizon of a black hole, thereby providing greater understanding of the general theory of relativity and the behavior of matter in this environment.

Finding: Radio wave bands (10 MHz to 3 THz) are indispensable for collecting information associated with specific astronomical phenomena. Often the same bands are similarly indispensable for passive Earth remote sensing, and the passive nature of both services enables them to productively share the spectrum.

3.4 SENSITIVITY REQUIREMENTS

Just as do Earth remote sensing researchers (see Chapter 2), radio astronomers use microwave radiometers to measure the total noise power received when an RAS telescope is pointed in a particular direction. The power received from an astronomical source is usually much less than that generated in the amplifiers and electronics or than stray radiation picked up from the ground (which emits at about 300 K). Radiometers and telescopes are carefully designed to minimize this contaminating signal and to keep it stable so that it can be subtracted to find the signal of interest. Radiometers may be broadband (with bandwidths from 10s of megahertz to several gigahertz) to maximize sensitivity to continuum sources, or they may be optimized for spectral-line observations, using a spectrometer that divides the radiation received over a broadband into many thousands of narrow channels. Radiometers may also be designed to be sensitive to the linear or circular polarization of the received radiation, which carries additional important information such as the direction of the magnetic field in a synchrotron-emitting plasma: radio astronomy is a very versatile and powerful probe of astronomical magnetic fields.

The power radiated by an extended source at frequency f is usually expressed as a "brightness temperature" T_b, which is the temperature of a blackbody that would emit the same amount of radiation, at that frequency. Brightness tempera-

[12]S.S. Doeleman et al., "Event-Horizon-Scale Structure in the Supermassive Black Hole Candidate at the Galactic Centre," *Nature*, 455: 78-80 (September 2008).

tures range from 2.7 K for the CMB to more than 10^{12} K for energetic nonthermal sources (pulsars, masers, and quasars). Astronomers, however, are often interested in much smaller differences of brightness temperature: for example, the tiny variations in the CMB temperature from one direction to another, which are only a few microkelvin.

The radio power incident on the antenna is called the flux density and is usually denoted by S_f; it is also called the spectral power flux density or spectral pfd. Flux density is measured in janskys (Jy, named after the pioneer radio astronomer Karl Jansky), where 1 Jy = 10^{-26} W m^{-2} Hz^{-1}. To bring the small magnitude of a jansky "down to Earth," consider that a garage-door opener on the Moon would produce about 5 Jy on Earth. A television transmitter on the planet Jupiter, more than 600 million kilometers away, would produce about 1 Jy on Earth. As another example, the Sun is a nearly ideal blackbody with temperature T ≈ 5,800 K. On Earth, its flux density at 10 GHz is about 1.2×10^6 Jy. By the inverse-square law, flux density decreases with distance as r^{-2}. Currently, the weakest detectable cosmic radio sources have flux densities about 1 microjansky, so a star like the Sun could be detected out to a million AU, or about one-tenth of a light-year. This is substantially less than the distance to the nearest star. Although the thermal radio emission from stars like the Sun cannot be detected at great distances, the thermal radio emission from other types of stars can be detected. More importantly, the more luminous sources in the universe, including quasars and gamma-ray bursts, can be detected at redshifts of 5 or more, corresponding to 90 percent of the way across the universe, or to the time when galaxies were first condensing from the primordial universe.

The signal-to-noise ratio (SNR) with which a source can be detected is approximately given by

$$\mathrm{SNR} = S_f A_{eff} (B\,\tau)^{1/2} / 2\mathrm{k}\, T_{sys}$$

where k is Boltzmann's constant (1.38×10^{-23} W Hz^{-1} K^{-1}), B is the bandwidth of the radiometer, τ is the integration time, and T_{sys} is the system temperature, a measure of the radiometer noise. The SNR generally must be 3 or greater for a positive detection, but a statistically sound result usually requires SNR = 5 or more.

The system temperature T_{sys} expresses the total unwanted noise power entering the receiver or generated in it as an equivalent temperature and is measured in degrees kelvin; it includes contributions from the sky (including the CMB and emission from the Milky Way), from the atmosphere, from the ground around the telescope, from interference (RFI), and from the telescope and amplifiers. The relative strength of these components and their absolute magnitude vary widely with frequency. Except for solar bursts, the signal from the source under study is usually much smaller than the system noise.

Sensitivity Limits

Radio astronomers maximize the SNR in their studies by using antennas with large collecting areas, such as the 100 m Green Bank Telescope or the 305 m Arecibo dish. They now have a project, the Square Kilometer Array, that will have a collecting area of about 1 km^2, or 10^6 m^2. The SKA is in the study and prototype phase; an optimum location for it should be selected within a few years' time.

Using the widest possible bandwidth also maximizes the SNR for a continuum source. Some modern radiometers have fractional bandwidths as wide as $\Delta f/f = 20\%$ or more. Broadband observations are essential for detecting the most distant known galaxies and the tiny fluctuations in the brightness of the CMB. The usable bandwidth is often limited by RFI.

Radio astronomers generally use very long integration times (hours or days) to maximize the SNR, but for some observations (pulsars and other transients), the integration time is limited by the duration of the signal itself. This can be a millisecond or less. The SNR can also be increased by reducing the system temperature, but current technology is close to the minimum possible T_{sys} for frequencies less than 100 GHz. The sensitivity of existing and proposed telescopes is shown in Figure 3.9.

A simple example will illustrate these ideas. Let the GBT look at a 10 microjansky source. Then the received spectral power is 4×10^{-28} watts per hertz, or, in common engineering units, –244 dBm Hz^{-1}. If the receiver has a system temperature of 30 K, then the noise spectral power is 4×10^{-22} W Hz^{-1}, or –184 dBm Hz^{-1}, a million times larger than the signal power. To make a positive detection by smoothing the receiver output to SNR = 3 requires that the product $B\tau$ be 10^{12}. This could be obtained, for example, with $B = 100$ MHz and $\tau = 10^4$ seconds. If a measurement of the flux density to 10 percent accuracy is wanted on this source, then the product $B\tau$ must be increased by a factor of 100, requiring a bandwidth of 1,000 MHz and an integration time of 10^5 seconds, longer than a day.

Observations like this are already being done at gigahertz frequencies, and they will become more common as new broadband instrumentation spreads throughout the radio community (Figure 3.9). Such observations are passive and cause no interference, but they use much more spectrum than is allocated to the RAS. The RAS bands, however, still are important for many narrowband observations that also are routinely done—for example, on spectral lines—and they are vital for the EESS. Note, however, that some extremely important astrophysical problems, such as studying redshifted HI with the 1420 MHz spectral line of HI, will need the entire range from 1420 MHz down to about 100 MHz. This broadband passive use by the RAS means that RFI outside the protected bands is of serious concern. It drives the observatory locations to remote sites such as Western Australia and will force consideration of the back side of the Moon as a possible radio observatory

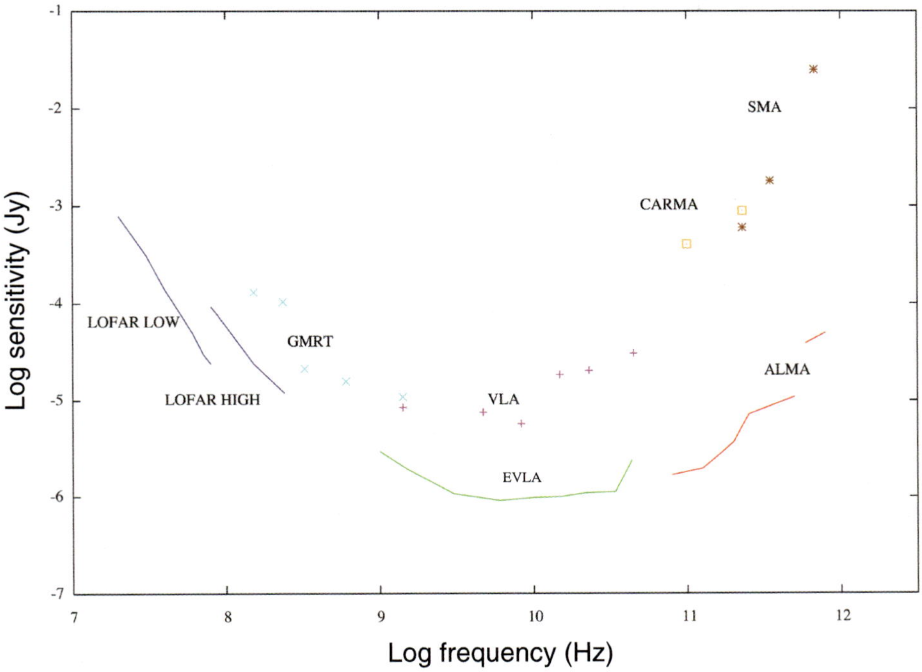

FIGURE 3.9 The root-mean-square sensitivity of various high-angular-resolution arrays in radio astronomy as a function of frequency. The discrete symbols refer to instruments that are in operation now. They are generally tunable by about +/-20 percent of their indicated frequencies. The solid lines refer to instruments that are under construction and will be operational by 2012. Note that the instruments under construction are between one and three orders of magnitude more sensitive than the existing ones. The sensitivity is proportional to the system temperature of the receivers and inversely proportional to the collecting area and the square root of the bandwidth and integration time, which in all cases is taken as 12 hours. The sensitivities were calculated from the array specifications on the Web sites of each instrument. LOFAR = Low Frequency Array (Netherlands); GMRT = Giant Metrewave Radio Telescope (India); EVLA = Expanded Very Large Array (New Mexico); VLA = Very Large Array (New Mexico); ALMA = Atacama Large Millimeter Array (Chile); CARMA = Combined Array for Research in Millimeter-wave Astronomy (California); SMA = Submillimeter Array (Hawaii). See Table 3.1 in this chapter for more information. Image courtesy of James Moran, Harvard-Smithsonian Center for Astrophysics.

site. This RFI is also a strong driver for the development of mitigation studies and technologies within the RAS community.

3.5 INTERFERENCE AND ITS MITIGATION

Radio astronomy deals with exceedingly weak signals. As described in §3.4, they can be a million times smaller than the internal receiver noise, and their measurement, or even just their detection, can require bandwidths of many gigahertz and integration times of a day or more. This requirement puts a premium on operating in a very low noise environment. It should be emphasized that serious interference can result from weak transmitters even when they are situated in the sidelobes of a radio astronomy antenna. This state of affairs has been recognized by the International Telecommunication Union (ITU) internationally and by the FCC in the United States, and various spectral bands have been allocated to the RAS for "exclusive" or "shared" use of these bands. However, "exclusive" does not mean that there must be zero emission in the protected bands. It is a fundamental fact that any information-carrying signal must contain out-of-band emission, which spreads across a wide radio spectrum. The regulation of this necessary out-of-band emission from a licensed transmitter involves controlling the intensity of the emission, and the FCC definition leads to an allowable level that, unfortunately, can cause serious interference with radio astronomy observations. It is likely that this situation will become worse in the future, as the RAS requirements become stricter with the study of weaker sources, and at the same time the active services are proliferating.

ITU-R Recommendation RA.769 discusses interference protection criteria for the Radio Astronomy Service and defines threshold levels of emissions that cause interference detrimental to radio astronomy. However, for modern measurements these levels are unrealistic, because they are not based on the current state of the art. The levels are calculated as 10 percent of the noise fluctuations, but the noise is calculated with a bandwidth of the allocated channel. However, bandwidths hundreds of times wider than this are routinely used. In fact, much of radio astronomy would no longer be possible if observations were restricted to the allocated channels. The other factor in the noise calculation, the integration time, is assumed to be 2,000 seconds, whereas in modern practice the integration times often are 10 or 50 times longer. Again, if observations were limited to 2,000 seconds, much of radio astronomy, especially the new realms projected for the coming decade, would be impossible. Hence, the limits set by ITU-R Recommendation RA.769 are inadequate today, and they will become more so in the future. This means that unwanted emissions that are legal can be damaging to the RAS measurements.

Another facet of the interference problem comes from emissions that essentially are unregulated. Cordless telephones, garage-door openers, and other unlicensed devices are allowed to have some low level of emissions, and at radio observatories an attempt is made to restrict the use of such consumer devices. But in fact they are powerful by RAS standards, as seen by the example of the garage-door opener on the Moon in §3.4, and will cause serious RFI if they are in the near sidelobes of a large antenna, even if they are far away. This problem also is worsening with the availability of new devices and their more widespread use. The incipient widespread use of automotive anticollision radar, operating at K-band, is a cause for concern in this regard.

A further cause of harmful RFI comes from transmitters that are operating illegally, either by producing excessive spurious or out-of-band transmissions or by operating at an unassigned frequency. The 1400-1427 MHz band is allocated to the RAS on an exclusive primary basis, but strong RFI has been seen in this band at many radio observatories around the world. Better monitoring of the radio spectrum and allocations would provide a better understanding of actual interference levels.

Radio observatories are located in remote sites, often behind mountains, to reduce human-made noise, which is roughly proportional to the local population density. But the problem is particularly severe with aircraft and satellite transmissions, from which there is no escape. Observations of transient phenomena are especially vulnerable to RFI because of the highly variable nature of both the phenomenon and the RFI.

Finding: The rules for out-of-band and spurious emissions in the primary allocated Radio Astronomy Service (RAS) bands (e.g., 1400-1427 MHz) do not provide adequate interference protection for RAS purposes.

The FCC rules that pertain to the above finding are given in Appendix D.

Finding: Geographical separation of radio telescopes from transmitters (e.g., through the establishment of radio quiet zones and the remote siting of observatories) is currently effective in avoiding much radio frequency interference, but the proliferation of airborne and satellite transmissions and the widespread deployment of mobile, low-power personal devices threaten even the most remote sites.

Examples of Interference in a Protected Band

Figure 3.10 shows interference in the band 1610.6-1613.8 MHz, which is allocated to the RAS on a shared primary basis—the interference was received in a 12 m antenna when an Iridium satellite passed through the beam. The satellite operates in the Mobile Satellite Service (MSS) band 1618.25-1626.5 MHz and, as seen in the figure, emits spurious radiation at 1612 MHz. During the measurement for Figure 3.10, careful attention was paid to ensure that the radiation was from the Iridium satellite itself and not from a GLONASS satellite, and that the RFI was not due to intermodulation in the receiver. Figure 3.11 shows the effect of similar satellite interference on an image made with the VLA in the same protected band, 1610.6-1613.8 MHz. The image made in the presence of the RFI is useless.

The 1610.6-1613.8 MHz band is most commonly used by radio astronomers for studying the OH radical that exists in stellar atmospheres and in clouds in the Milky Way; the studies are conducted in a spectroscopic mode in which many

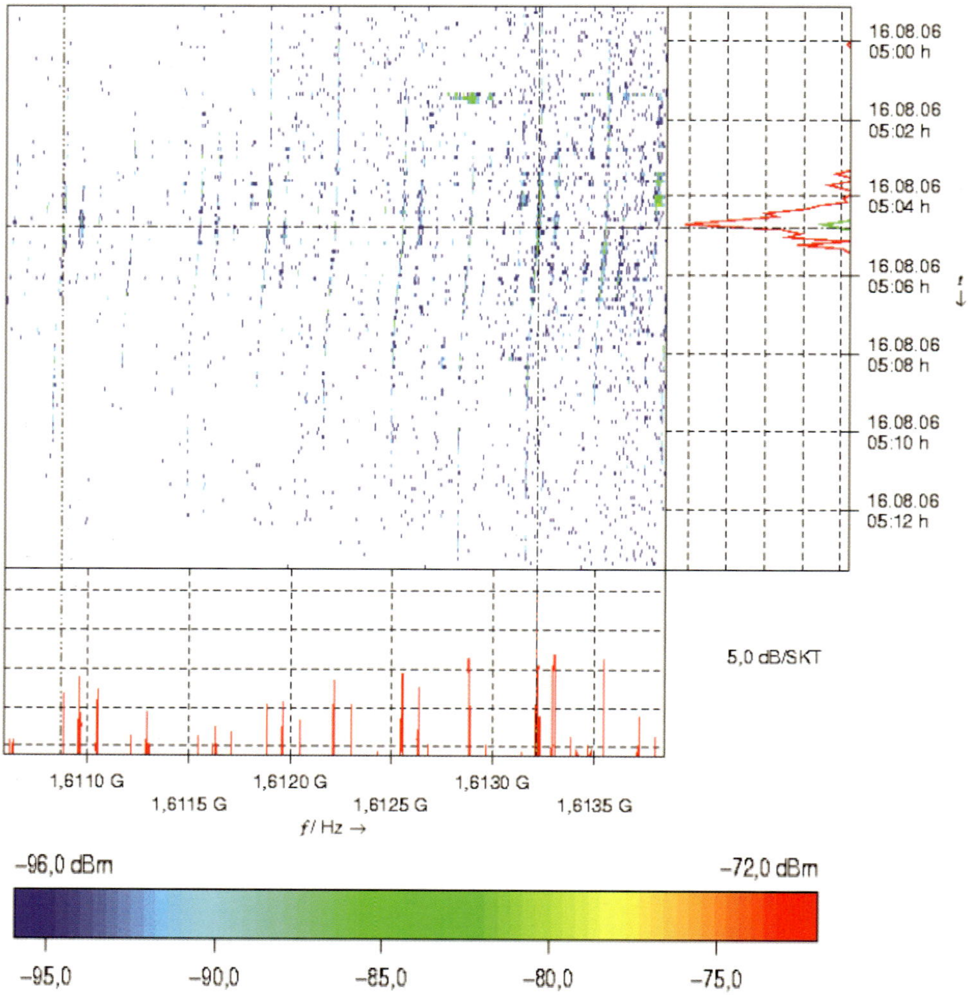

FIGURE 3.10 Showing radio frequency interference due to spurious emission from an Iridium satellite in the band 1610.6-1613.8 MHz, which is allocated to the Radio Astronomy Service on a primary basis. This measurement was made in Leeheim, Germany, in November 2006, with a 12 meter parabolic antenna. Careful attention was paid to eliminating the possibility of unwanted interference from intermodulation products in the receiver. Time runs down in the graph, over a total of 14 minutes, and frequency is horizontal. The motion of the satellite can be seen in the changing Doppler shift of the signals as the satellite passes through the beam of the antenna. The peak is about –85 dBm, substantially higher than the value recommended by the International Telecommunication Union, when it is converted to the standard model using an isotropic antenna. When converted to standard radio astronomy units, the flux density during the short bursts is about 2,500 Jy. Image courtesy of CEPT and BNetzA.

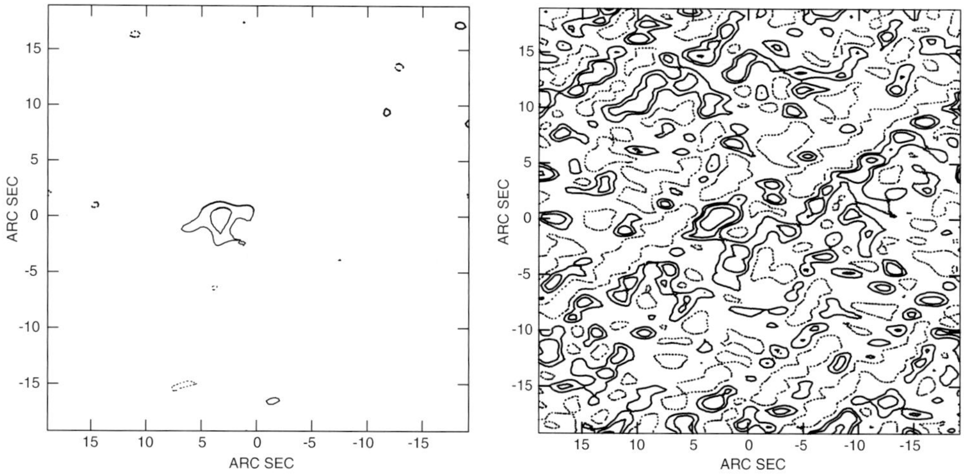

FIGURE 3.11 The effect of radio frequency interference (RFI) on an astronomical image made at the Very Large Array. (Left) Image of a faint "OH/IR star" made in a narrow band at 1612.22 MHz, within the band 1610.6-1613.8 MHz that is allocated to the Radio Astronomy Service on a primary basis. (Right) The same field of observation of the image made when an Iridium satellite was 22 degrees from the star. This image is made useless by the RFI. Images courtesy of G.B. Taylor, University of New Mexico.

narrow bands are measured simultaneously. The RFI depicted in Figure 3.10 could adversely affect OH observations made when the satellite is well outside the main beam of the antenna, even for a large antenna like the GBT that has a forward gain of 63 dBi at 1612 MHz. The potential for harmful RFI is high, especially considering that the Iridium constellation contains 66 satellites.

Mitigation

"Unwanted" emission is of two kinds: "out-of-band" and "spurious." Out-of-band emission is unwanted emission on a frequency or frequencies immediately outside the transmitter's necessary bandwidth; it results from the modulation process. This type of emission is different from spurious emission, which results from harmonics or intermodulation products generated in the transmitter. When considering the regulation of signals that may spill into science service bands, account should be taken of how such signals will affect the scientific instruments in question.

Simple excision techniques, in both time and frequency, have long been used to mitigate the effects of interference. More sophisticated procedures using statistical methods are currently under investigation, as described in Chapter 4. The present period is one of increasing sensitivity in the radio astronomy systems and increasing use of the spectrum by other users, particularly the low-power wireless applications. These needs are conflicting, and the interference problem will undoubtedly increase. Focusing more attention on mitigation possibilities is important. At the same time, the radio astronomy enterprise must be protected by increased vigilance over its protected bands.

The approach to reducing the impact of RFI at radio observatories occurs at several different levels, depending on the resources at each observatory. These approaches are briefly described in the following subsections. See Chapter 4 for additional discussion on this topic.

Regulatory and International Meetings

Only the largest observatories (e.g., the National Astronomy and Ionosphere Center [NAIC] and the National Radio Astronomy Observatory [NRAO]) are normally able to provide continuous staff attendance at international meetings, such as regular ITU Working Party 7D (WP7D) meetings and the World Radiocommunication Conference (WRC). However, smaller, university observatories are kept informed of events in the international arena by regular teleconferences among the observatories, and by attendance at the U.S. WP7D teleconferences. The NRAO has a spectrum manager, an astronomer who pursues his own astronomical research but who spends a significant fraction of his time on spectrum management activities, including responding to the FCC on NRAO's behalf and contributing to and attending international ITU meetings.

Quiet Zones Around Observatories

Only two observatories on U.S. soil benefit from Quiet Zone protection: NRAO (Green Bank, West Virginia) and NAIC (Arecibo, Puerto Rico). In addition, the United States is a major partner in the ALMA project being built in northern Chile. The Chilean authorities, through the Subsecretaría de Telecomunicaciones (SUBTEL), have agreed to a considerable level of protection from interference from other services around the ALMA site.

The administration of these Quiet Zones requires resources. For example, in the case of the National Radio Quiet Zone (NRQZ) at Green Bank, West Virginia, all applications for fixed transmitters within the NRQZ are examined by NRAO staff, who make comments to the FCC on the basis of a technical analysis, usually including propagation predictions over the specific path. Often, some compromise as to power, frequency, and, in particular, precise location of the new transmitter, is agreed to between the parties concerned.

The administration of an NRQZ by a radio observatory requires a significant, continuing effort. However, this effort is usually very well rewarded. For example, at the Green Bank observatory in West Virginia during 2007, 538 requests for coordination within the Quiet Zone were processed. They involved 850 sites within the Quiet Zone and 872 transmission frequencies. In 13 cases, a site inspection was carried out. For about a dozen of the requests, a power restriction was eventually placed on the applicant's FCC transmitter license. However, in a far greater number of cases, a solution agreeable to both parties, one that did not necessarily restrict the transmitter power, was negotiated. The negotiations usually resulted in alternative transmitter sites and/or directional antennas pointed away from the observatory, with a compromise in capability for the transmitter operator, while still providing adequate protection for the observatory.

Local Radio Frequency Interference

The NRAO engineering staff at Green Bank includes a team that tracks down instances of RFI that appear at the observatory. The team's equipment includes a portable interference system, which can trace interference originating within a few miles of the observatory. If it is possible technically to suppress the interfering source, by simple technical means or perhaps by negotiation with the relevant party, this is done. In very rare cases, where the aforementioned methods fail, the FCC may be called on to intercede.

Local Engineering

The observatories themselves take all practical engineering precautions in the design and construction of equipment in order to provide adequate filtering and dynamic range so as to make equipment as immune as possible to interference from out-of-band signals. Special techniques are sometimes used, such as a dedicated antenna to monitor a particular source of interference; the data gathered by such monitoring can then be subtracted from the astronomical data by some means. This is more a research than an operational area at present, with few such systems currently in use. NRAO, for example, is investigating several mitigation possibilities, including active RFI cancellation, ways of extending dynamic range, and high-performance filtering using the latest technology.

Data Processing

RFI mitigation using software processing techniques is in routine use at most observatories. This approach includes data excision based on time or frequency; it is often carried out automatically, with some manual input. Other techniques are active research areas at a number of observatories, as described elsewhere in this report. See Chapter 4, in particular.

Finding: While unilateral radio frequency interference mitigation techniques are a potentially valuable means of facilitating spectrum sharing, they are not a substitute for primary allocated passive spectrum and the enforcement of regulations.

Finding: Important scientific inquiry and applications enabled by the Radio Astronomy Service (RAS) are significantly impeded or precluded by radio frequency interference (RFI). Such RFI has reduced the societal and scientific return of RAS observatories and necessitates costly interference mitigation, which is often insufficient to prevent RFI damage.

3.6 IMPORTANCE OF RADIO ASTRONOMY TO THE NATION

The science of radio astronomy started in 1932, with the accidental discovery of radio waves from the Milky Way by Karl Jansky. Little happened in this field during the 1930s, but during World War II the United States mobilized a huge development effort in radar technology. The instrumentation and techniques resulting from this work fueled modern research in radio astronomy.[13] Since then radio astronomy has continuously benefited from new technological developments; many of these have come from government and commercial sources, but some have been forthcoming from the development laboratories within radio astronomy itself. This section outlines some of the important benefits to the nation provided by radio astronomy.

Radio Interferometry

The development of interferometry has had widespread applications in fields in addition to radio astronomy and provides attendant benefits to society. The underlying principle of interferometry is the measurement of the relative time of arrival of signals from a radio source, among a group of antennas called an array. Triangulation then gives the direction of arrival of the radiation, meaning that the angular position of the radio source can be measured precisely. Furthermore, comparison of the arriving signals provides a method of imaging the source—that is, of determining the angular structure of the emission, which reveals the structure and dynamics of the source. These two applications—precise positioning and imaging—are important, as noted above, in fields beyond radio astronomy. For example, the radio technique of combining observations from different configurations of an array of antennas has formed the underlying principle of back projections, which is mathematically very closely related to all the techniques of medical imaging, such at computed axial tomography (CAT) and magnetic reso-

[13]R. Buderi, "The Invention That Changed the World: How a Small Group of Radar Pioneers Won the Second World War and Launched a Technological Revolution," *Touchstone* (March 1998).

nance imaging (MRI). There has been much cross-fertilization in the development of these techniques.[14]

The highest angular precision by radio interferometers has been achieved through the use of networks of telescopes distributed around the world and linked through a technique called very long baseline interferometry. To provide the time-of-arrival information, receiving stations must be equipped with precise clocks. The best technology for this purpose is the hydrogen maser frequency standard, developed originally at Harvard University and since perfected primarily for VLBI, radar astronomy, and space tracking needs. The positions of a set of very distant radio sources have been determined with VLBI networks, providing a stable, precise reference frame for a wide variety of applications. For example, with this established reference frame the relative motions of antennas on Earth can be tracked to an accuracy of a few millimeters per year. This capability led to the first measurement of the contemporary motions of tectonic plates. Fluctuations in the rotation rate of Earth and the orientation of its spin axis are continuously monitored this way and provide information useful to the understanding of the composition and motions of Earth's molten core and the annual changes in polar-ice loading.

These techniques of radio surveying based on triangulation formed the intellectual and technical basis for the development of the GPS and other terrestrial navigation systems. In the radio astronomy case, a distant radio source acts as a transmitter whose signal is received by a number of antennas, so that its position can be determined by relative time-of-arrival methods. In the GPS case, a user at an unknown location on Earth receives signals from an array of satellites, from which the user finds his or her position through triangulation based on a similar time-of-arrival analysis. GPS thus is highly analogous to the earlier VLBI, even to their both using atomic frequency standards as clocks.

Communications Disruptions

Energetic particles from the Sun, released in bursts called coronal mass ejections, arrive at Earth and can cause a disruption in radio communications, interference with GPS operation, surges on power grids, damage to Earth-orbiting satellites, and hazards to astronauts. The prediction of such events is important so that measures can be taken to ameliorate their effects. Amelioration, for example, might be achieved by shifting communications to less-affected frequencies and by placing satellites in standby mode. Hence the advance knowledge of the onset of these disruptive events is beneficial, just as the prediction of the arrival of meteorological events is important to the reduction of property damage and loss of life.

[14]National Research Council, *The Decade of Discovery in Astronomy and Astrophysics*, Washington, D.C.: National Academy Press, 1991, pp. 129-130.

Coronal mass ejections originate from disturbances on the Sun, generally in the form of prominences and flares. Because CMEs consist of charged particles, they continuously emit radio emission as they travel outward from the Sun, and they can be tracked by radio telescopes, allowing 1 to 2 days of warning. Flares often are associated with huge bursts of radio emission, which have been known to seriously interfere with GPS operations, as described in §3.1. One of the goals of solar physics is to make long-term predictions of flares and CMEs by studying the emissions from the Sun (see Figure 3.7).

Fundamental Physics

The recent discovery, based on astronomical observations, that normal matter (baryonic matter) constitutes only 4 percent of the mass of the universe, while the rest is in the form of dark matter and dark energy, is transforming the understanding of physics. Radio astronomical observations of the rotation of galaxies have proved to be an excellent way to trace the distribution of dark matter. Meanwhile, laboratory experiments are underway in an attempt to identify the particle nature of dark matter. This combined effort in astronomy and laboratory physics can be expected to lead to a major step forward in the understanding of the universe.

The measurement by radio astronomers of the timing of the rotations of pulsars in tight binary orbits about companion neutron stars, with exquisite precision, has been providing physicists with the strongest affirmative answer yet to the century-long question, "Was Einstein right?" (see Figure 3.3).

Technology Development

Radio astronomy has advanced the limits of technology as it has opened up spectral bands at progressively higher frequencies. For example, the best technology for low-noise receivers at frequencies above 100 GHz and into the terahertz range is based on quantum devices known as superconductor-insulator-superconductor (SIS) mixers. These devices were first developed by radio astronomers at the University of California in the 1980s and independently at AT&T Bell Laboratories. They have now been perfected, primarily for use in radio astronomy, to operate with noise levels at a few times the quantum limit. As military and telecommunications applications move into this band, they will undoubtedly make use of this technology.

Precision Antennas

The need for high sensitivity has led radio astronomers to develop the technology of building highly efficient, large, parabolic antennas, which have extensive application in the telecommunications and military communities. Radio astronomers first developed the theory of how to design large, fully steerable antennas that maintain high surface accuracy in the presence of gravitational deformations. They invented an electronic surveying technique, known colloquially as radio holography, which enables reflector surfaces to be set to an accuracy of a few microns. Methods that they developed for measuring antenna efficiency from observations of standard radio sources and solar system bodies are in wide use.

Distributed Network Computing

The Search for Extraterrestrial Intelligence (SETI) project, a search carried out in radio bands, was faced with an enormous computational problem in analyzing its voluminous data to find nonrandom signals that might be of extraterrestrial origin. The computing resources needed to sort through the collected data were far beyond those available to the SETI researchers. The solution was to enlist the aid of interested people, who would download an analysis program and a section of data and would do the analysis in their computer's background. More than 5 million people in 226 countries responded and are part of the SETI@home project. The SETI@home researchers went on to develop the Berkeley Open Infrastructure for Network Computing (BOINC). BOINC's open-source volunteer computing platform currently engages the public in 42 scientific supercomputing projects, including climate modeling and global warming studies (ClimatePrediction.net); drug research for HIV, malaria and cancer, and protein folding (Predictor@home); gravity waves (Einstein@home); particle physics (LHC@home); as well as SETI@home. BOINC volunteers provide about 2 petaflops of computing power to the various projects, more than the world's most powerful supercomputer.

Education and Public Outreach

Radio astronomy requires a broad spectrum of technically trained people, from theorists and observers with doctoral degrees to technicians with much less education, perhaps even trained primarily on the job. Theory, observations, and analysis are usually done by small teams consisting of one person or several senior people along with junior people, students, and postdoctoral researchers. Only a small fraction of the people trained for this field actually stay in radio astronomy; the majority go into other fields, usually still in a technical capacity. They form a valuable pool of people with a wide range of skills who readily find technical jobs in industry or government laboratories.

The general public is greatly interested in astronomy, perhaps more than in other sciences. There is a steady stream of astronomy stories and images in the press. At the nation's colleges and universities, it is the most common subject taken as a science requirement, with thousands of students per year in elementary astronomy classes at the larger universities. All the major radio observatories have well-attended visitor programs, with the Arecibo Observatory in Puerto Rico drawing 120,000 visitors annually, of whom 30 percent are children. This interest translates into an appreciation of science and technology, and draws students into technical subjects, helping to provide the personnel resources needed in today's world.

Finding: In addition to the intellectual benefits that they provide, radio astronomy studies provide many technological benefits to American society.

Finding: Radio astronomy provides a diverse and valuable set of educational opportunities.

4

Technology and Opportunities for the Mitigation of Radio Frequency Interference

The capacity to address interference issues is a key element in any system using the radio frequency (RF) spectrum. This is especially true for passive sensing systems—those that do not transmit but only receive naturally occurring emissions—due to the level of sensitivity required for these systems to extract useful environmental and scientific data. Interference can be caused by a variety of sources: other valid users of the RF spectrum, improperly functioning consumer and commercial equipment, and improper or disallowed use of the spectrum. As the use of the RF spectrum for commercial, industrial, government, and scientific uses continues to increase, the number of potentially interfering sources will increase as well. Mitigation techniques are a limited but critical element in efforts aimed at extracting scientific value from an increasingly difficult RF environment.

The Earth Exploration-Satellite Service (EESS) and Radio Astronomy Service (RAS) have classically limited the impact of interference by using mitigation techniques. However, there are physical limits to the capacity of the "unilateral" techniques that typically have been used, and they often do not provide adequate protection from interference. Recently, new techniques have been suggested, in which the active and passive users of the RF spectrum collaborate in order to share the spectrum. These "cooperative" mitigation techniques may provide a potential for meeting the expanding spectral needs of the passive-sensing community.

This chapter is divided into five sections: §4.1 addresses the expected trends in RF spectrum use that may call for increasing mitigation; §4.2, the drivers of spectrum use; §4.3, the capacity for unilateral mitigation technology; §4.4, the potential for mitigation through cooperative use; and §4.5, the costs of mitigation.

4.1 TRENDS IN ACTIVE SPECTRUM USAGE

One of the primary concerns for passive sensing systems is the explosive growth in industrial, commercial, and consumer devices. This growth is fueled by user demand, investment capital, and the reallocation of underutilized spectral bands. The need for mitigation and the appropriate mitigation technique will vary depending on the type of equipment that will be permitted by the regulatory agencies, the technology being deployed, the time line for the deployment of systems, and the intensity of spectrum usage. This section and the next present a review of current spectrum usage and of the drivers for future spectrum usage, which provides the requisite basis for the development of the appropriate technical and regulatory mitigation strategies.

Current Allocations

Access to spectrum in the United States is assigned by the Federal Communications Commission (FCC) and the National Telecommunications and Information Administration (NTIA). The process is described in useful detail in Chapter 1 of the National Research Council's 2007 report *Handbook of Frequency Allocations and Spectrum Protection for Scientific Uses.*[1] To summarize, spectrum is typically assigned to *services* (classes of users) on a primary basis or a secondary basis, and allocations include details on permitted transmission power levels and operation times. The difference between a primary allocation and a secondary allocation is essentially that the users of a secondary allocation must accept interference from the users of a primary allocation and conversely must not interfere with the users of the primary service. The International Telecommunication Union (ITU), an agency of the United Nations, periodically updates its allocation table to coordinate international spectrum usage and prevent problems due to interference. The ITU Radio Regulations (ITU-RR) are not binding on the United States in toto—the real treaty obligation of the U.S. government is that it will not assign transmitter licenses in such a way that will cause interference to stations licensed by other governments that are in accordance with the ITU-RR. Within this framework, national governments create and enforce additional regulations, typically to include additional details and to elaborate on permitted uses of the spectrum. In the United States, federal use of spectrum is managed by the NTIA, whereas nonfederal (i.e., commercial, amateur, and passive scientific) use of spectrum is managed by the FCC. The authority of the FCC and NTIA are parallel in this respect. FCC regulations

[1] National Research Council, *Handbook of Frequency Allocations and Spectrum Protection for Scientific Uses*, Washington, D.C.: The National Academies Press, 2007.

concerning the use of the spectrum are codified in Title 47 of the *Code of Federal Regulations* (47 C.F.R.).

The radio astronomy community is represented in this process as the "Radio Astronomy Service," and the Earth remote sensing community is represented in this process as the "Earth Exploration-Satellite Service." A useful synopsis of 47 C.F.R. in terms relevant to the RAS and EESS, including tables of relevant spectral allocations, is given in the *Handbook*.[2] For example, this reference text shows that 2.07 percent of the spectrum below 3 GHz is allocated to the RAS and EESS on a primary basis and that 4.08 percent is allocated on a secondary basis (measured in hertz).

From a regulatory perspective, the RAS and EESS are comparable to all other services, despite the fact that that they do not transmit. Thus, the allocation of spectrum to the RAS and EESS on a primary basis results nominally (but not actually—see below) in clear spectrum. The allocation of spectrum to the RAS and EESS on a secondary basis is useful mainly in the sense that it offers these services a legal basis for providing input into the use of these allocations. It should also be noted that the allocation of a frequency band to the RAS and/or EESS does not prevent interference even if the allocation is on a primary basis. This is because the effective bandwidth of any transmission is essentially unlimited when observed with a sufficiently sensitive instrument. So, for example, the far out-of-band (sideband) emission of a transmission whose center frequency is properly in a band in which it has a primary allocation may, at some level, appear in nearby bands in which the RAS and/or EESS has a primary allocation. This has historically been a severe problem, particularly with respect to interference from services transmitting from satellites in L-band. (See §3.5 for a discussion of radio frequency interference [RFI] from Iridium satellites.) In contrast to active uses of the spectrum, the work of RAS and EESS users can be severely affected when the interference power level is far below the internal noise power level of the detection device, since long integration times are usually used in RAS and EESS measurements to reduce the root-mean-square fluctuations in the internal noise. Thus, this issue affects the RAS and EESS in a way that is fundamentally different from the way that it affects active users of the radio spectrum.

The spectrum in which the RAS and/or EESS has a primary or secondary allocation is relatively small (see Table 4.1). The spectrum in which the RAS and/or EESS has a secondary allocation has diminished usefulness, since there is no protection from the primary users of these bands. As noted in Chapters 2 and 3, the spectrum requirements of the radio astronomy and Earth exploration radio science community currently far exceed the spectrum available to the RAS and EESS on

[2] National Research Council, *Handbook of Frequency Allocations and Spectrum Protection for Scientific Uses*, Washington, D.C.: The National Academies Press, 2007.

TABLE 4.1 Total Spectrum Allocated to the Radio Astronomy Service (RAS) and the Earth Exploration-Satellite Service (EESS) Within 9 kHz to 3 GHz

	Total Bandwidth Allocation (MHz)	Percent of Bandwidth Allocated (%)
EESS only (9 kHz-3 GHz)		
Primary	37	1.23
Secondary	122	4.07
RAS only (9 kHz-3 GHz)		
Primary	62.12	2.07
Secondary	35.5	1.18
RAS + EESS (9 kHz-3 GHz)		
Primary	62.12	2.07
Secondary	122.5	4.08

NOTE: The percentage and bandwidth allocated to the RAS and EESS between 9 kHz and 3 GHz as of this writing is given in the table above. Note that "RAS + EESS" is much less than the sum of RAS and EESS, particularly in primary bands, showing that the two services are able to share spectrum efficiently.

either a primary or a secondary basis. For this reason, these users must routinely observe in bands in which the RAS and/or EESS has neither primary nor secondary allocations. This is authorized since passive (receive-only) scientific use of the radio spectrum is not prohibited in any part of the electromagnetic spectrum. This is also often technically possible because some parts of the spectrum are sparsely utilized, and services that transmit typically do so with poor spectral efficiency in both frequency and time (although current trends are in the direction of increased usage and improved spectral efficiency; see §4.2).

Finding: Owing to their receive-only nature, the passive Earth Exploration-Satellite Service and Radio Astronomy Service, operating from 10 MHz to 3 THz, are incapable of interfering with other services.

Finding: Currently, 2.07 percent of the spectrum below 3 GHz is allocated to the RAS and EESS on a primary basis, and 4.08 percent is allocated on a secondary basis (measured in hertz).

Current Utilization Studies

The allocation of spectrum to a service does not necessarily imply that the allocated spectrum is always used for transmission; neither does it imply that the allocated spectrum cannot be used by others for passive scientific observations. In fact, there are a number of ways in which allocated spectrum might remain free of detectable transmissions and available for passive scientific observations. For example, in rural areas with low population density, some services may not be used to any detectable degree; that is, transmissions associated with the active services are typically fewer and weaker in these areas (see further discussion later in this section). If the area is remote enough, transmissions may be sufficiently weak so as not to interfere with radio science observations (although such remote areas are reached by satellite and airborne radio sources). More often, however, the situation is intermediate in the sense that significant interference is observed but can sometimes be managed through a combination of interference mitigation techniques (see §4.3 and §4.4).

The "channelization" of frequencies within a given allocation, typically specified either in 47 C.F.R. or as the result of the adoption of an industry standard (e.g., IEEE 802.11), is inherently inefficient. For example, a typical user of the Land Mobile Radio Service might use only a small number of widely spaced channels within the allocated band and may transmit on them only a tiny fraction of the time. Thus even if an active user is received with sufficient strength to prevent the scientific use of some section of the spectrum while that user is transmitting, it is sometimes possible to exploit the sparse "time-frequency" utilization of spectrum by means of the techniques described in §4.3 and §4.4 to observe effectively when the active user is not transmitting. Given the existing trend toward more efficient channelization and increased utilization, however, interference mitigation methods that rely on this property are in danger of becoming less effective over time.

The modulation employed by a transmitter may be inherently inefficient, in the sense that it requires a large swath of spectrum but unevenly distributes the power over the channel. An example is the use of the National Television System Committee (NTSC) standard for analog television (TV), which requires a 6 MHz channel but places the vast majority of the transmitted power into just two carriers constituting only a few hundred kilohertz of bandwidth within this channel (see Figure 4.1). Radio astronomers have been able to observe within active NTSC channels in areas where NTSC transmissions are relatively weak (e.g., deep in the National Radio Quiet Zone [NRQZ]) by observing only within those portions of the channel where relatively little power is present and filtering out those parts of the channel where most of the power is located. However, the introduction of the new digital TV broadcast standard, known as ATSC, makes this technique impossible. This is because the Advanced Television Systems Committee (ATSC) fills the entire 6 MHz channel with a uniform distribution of power, leaving no "hole" through which to observe (see Figure 4.1).

The preceding comments can be summarized as follows: (1) The "allocation" of spectrum historically has not implied the "utilization" of spectrum, which has benefited the passive scientific users of the radio spectrum. (2) Technology trends are moving toward more efficient utilization of allocations, in both time and frequency, which is beginning to severely impact the ability to use some bands

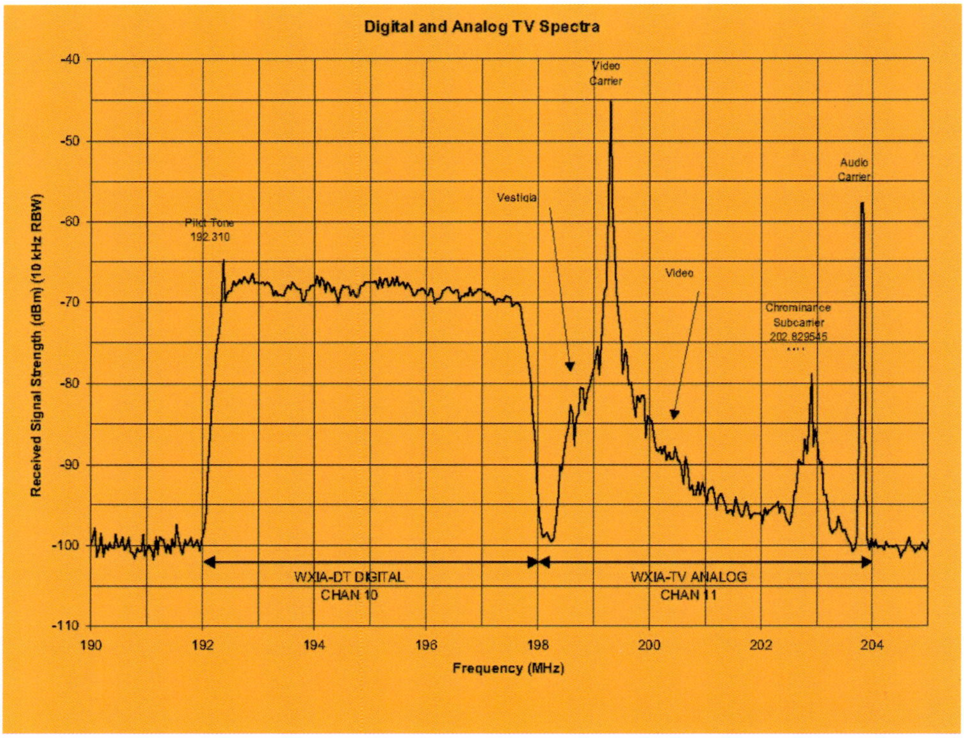

FIGURE 4.1 A comparison of the digital (Channel 10) and analog (Channel 11) television signals. The digital signal is essentially uniform in power across the entire channel, whereas the analog signal transmits most of its information in two narrow bands, leaving holes through which radio astronomers can sometimes observe relatively strong natural sources. Image courtesy of Andrew Clegg, National Science Foundation.

for scientific uses, despite the importance of these uses. As explained later in this chapter, this is true even when taking into account the capabilities of existing and emerging techniques for the mitigation of interference.

For these reasons, passive-sensing scientific users of the radio spectrum are greatly concerned with the utilization of spectrum both within the bands allocated to the RAS and EESS and in all other bands accessible to current and planned instruments. Furthermore, radio astronomers are concerned not only with spectral occupancy at frequencies at which they wish to observe, but they also monitor transmissions in nearby bands that have the potential to create interference through receiver compression—a condition in which an instrument is desensitized because a signal in a nearby band is present with such great strength that the receiver goes nonlinear. In this case the ability to mitigate the interference through filtering, while retaining sensitivity, is beyond existing technology. The reason for this is that filters must be placed before the saturable active components, and because filters have losses, the system sensitivity is thereby reduced.

The largest radio observatories routinely monitor the RF spectrum and typically maintain continuous monitoring programs of some sort. The results of monitoring campaigns are usually freely available for inspection online—for example, at the interference-monitoring Web sites maintained by the NRAO's Very Large Array (VLA),[3] by NRAO at the Green Bank Telescope,[4] and by the National Atmosphere and Ionosphere Center (NAIC) at Arecibo Observatory.[5] Unfortunately, these efforts are technically difficult and expensive to maintain and thus often have limited sensitivity and/or restricted time-frequency coverage. As a result, interference that is strong enough to be harmful to radio astronomy may escape detection by existing monitors. With regard to the EESS community, the monitoring of spectral utilization is made even more difficult by the limitations of operations aboard aircraft and satellites and by the coarse spectral resolution of total power radiometers. However, some anecdotal results have been published (see Figures 2.13 through 2.21, 3.10, and 4.2).[6]

The actual utilization of the radio frequency spectrum has recently become a topic of increasing interest to active users of the spectrum as well. This has resulted in a number of studies reporting measurements of the utilization of the spectrum.[7] Typically, the results of such studies report results in terms of *spectral occupancy,*

---

[3]See "VLA Radio Frequency Interference" at http://www.vla.nrao.edu/cgi-bin/rfi.cgi; accessed January 13, 2009.

[4]See "Green Bank Interference Protection Group" at http://www.gb.nrao.edu/IPG/; accessed January 13, 2009.

[5]See National Atmosphere and Ionosphere Center, Arecibo RFI Web site, at http://www.naic.edu/~rfiuser/; accessed January 13, 2009.

[6]S.W. Ellingson, G.A. Hampson, and J.T. Johnson, "Characterization of L-Band RFI and Implications for Mitigation Techniques," *Proc. IEEE Geoscience and Remote Sensing Symp. (IGARSS 2003)*, Vol. 3, pp. 1745-1747, July 21-25, 2003.

[7]Federal Communications Commission Spectrum Policy Task Force, *Report of the Spectrum Efficiency Working Group*, November 2002, available at http://www.fcc.gov/sptf/reports.html, accessed January 14, 2010; Frank H. Sanders and Vince S. Lawrence, *Broadband Spectrum Survey at Denver, Colorado*, NTIA Report 95-321, September 1995; Frank H. Sanders, Bradley J. Ramsey, and Vincent S. Lawrence, *Broadband Spectrum Survey at San Francisco, CA*, NTIA Report 99-367, May-June 1999; S.W. Ellingson, "Spectral Occupancy at VHF: Implications for Frequency-Agile Cognitive Radios," *Proc. IEEE Vehicular Technology Conf. 2005 Fall—Dallas*, Vol. 2, pp. 1379-1382 (September 2005); A.E.E. Rogers, J.E. Salah, D.L. Smythe, P. Pratap, J.C. Carter, and M. Derome, "Interference Temperature Measurements from 70 to 1500 MHz in Suburban and Rural Environments of the Northeast," *Proc. First Int'l Symp. on New Frontiers in Dynamic Spectrum Access Networks (DySPAN 2005)*, November 8-11, 2005, pp. 119-123; Mark A. McHenry and Dan McCloskey, "Multi-Band, Multi-Location Spectrum Occupancy Measurements," *Proc. 2006 ISART Conference, Boulder, Colorado* (March 2006), available at http://www.its.bldrdoc.gov/pub/ntia-rpt/06-438, accessed January 14, 2010.

which can be defined as the fraction of time that a transmission can be detected at a given frequency, for a given sensitivity and a given time-frequency resolution.

However, the perception of what constitutes occupancy can be different depending on the measurement and the interests of the person interpreting the results. For example, a recent study performed by the Shared Spectrum Corporation reported 13.1 percent occupancy for New York City and 1 percent at Green Bank, West Virginia, inside the NRQZ.[8] By contrast, a study of occupancies in terms somewhat more relevant to radio astronomy applications finds occupancy greater than 30 percent even in the relatively rural areas of Westford, Massachusetts, and Hancock, New Hampshire.[9] Both studies are probably internally consistent but cannot be compared because of their different assumptions about the appropriate time-frequency resolutions, thresholds of detection, and tolerable levels of out-of-band (OOB) interference. Measurements are also being made that attempt to bridge this gap by reporting results in terms of cumulative distribution functions (CDFs), which resolve "occupancy" as a function of threshold of detection, and by also quantifying the fragmentation of unoccupied bandwidth. This activity is important, because often in both active and passive uses of the spectrum a minimum bandwidth must be available for the channel to be useful.[10]

While the various efforts of the active and passive user communities have been useful in confirming the sparse time-frequency utilization of the spectrum, most existing studies are of limited help in understanding in detail the potential for interference and for cooperative spectrum use as described later in this chapter. This is due to limited sensitivity (i.e., the inability to detect weak signals that still are sufficiently strong to constitute "occupancy" to a typical user of that band), time resolution that is too coarse to be useful (for example, monitoring a frequency for only a few milliseconds every few seconds, thereby potentially missing strong signals), and frequency resolution that is too coarse to be useful (for example, monitoring bandwidths on the order of hundreds of kilohertz when the signals themselves have bandwidths on the order of kilohertz, thereby desensitizing the measurements) (see Figure 4.2). This is essentially the same problem experienced by the monitoring programs of radio observatories, as mentioned above. Thus,

---

[8] Mark A. McHenry and Dan McCloskey, "Multi-Band, Multi-Location Spectrum Occupancy Measurements," *Proc. 2006 ISART Conference, Boulder, Colorado* (March 2006), available at http://www.its.bldrdoc.gov/pub/ntia-rpt/06-438/; accessed January 14, 2010.

[9] A.E.E. Rogers, J.E. Salah, D.L. Smythe, P. Pratap, J.C. Carter, and M. Derome, "Interference Temperature Measurements from 70 to 1500 MHz in Suburban and Rural Environments of the Northeast," *Proc. First Int'l Symp. on New Frontiers in Dynamic Spectrum Access Networks (DySPAN 2005)*, November 8-11, 2005, pp. 119-123.

[10] S.W. Ellingson, "Spectral Occupancy at VHF: Implications for Frequency-Agile Cognitive Radios," *Proceedings of the IEEE Vehicular Technology Conference 2005 Fall—Dallas*, Vol. 2, pp. 1379-1382 (September 2005).

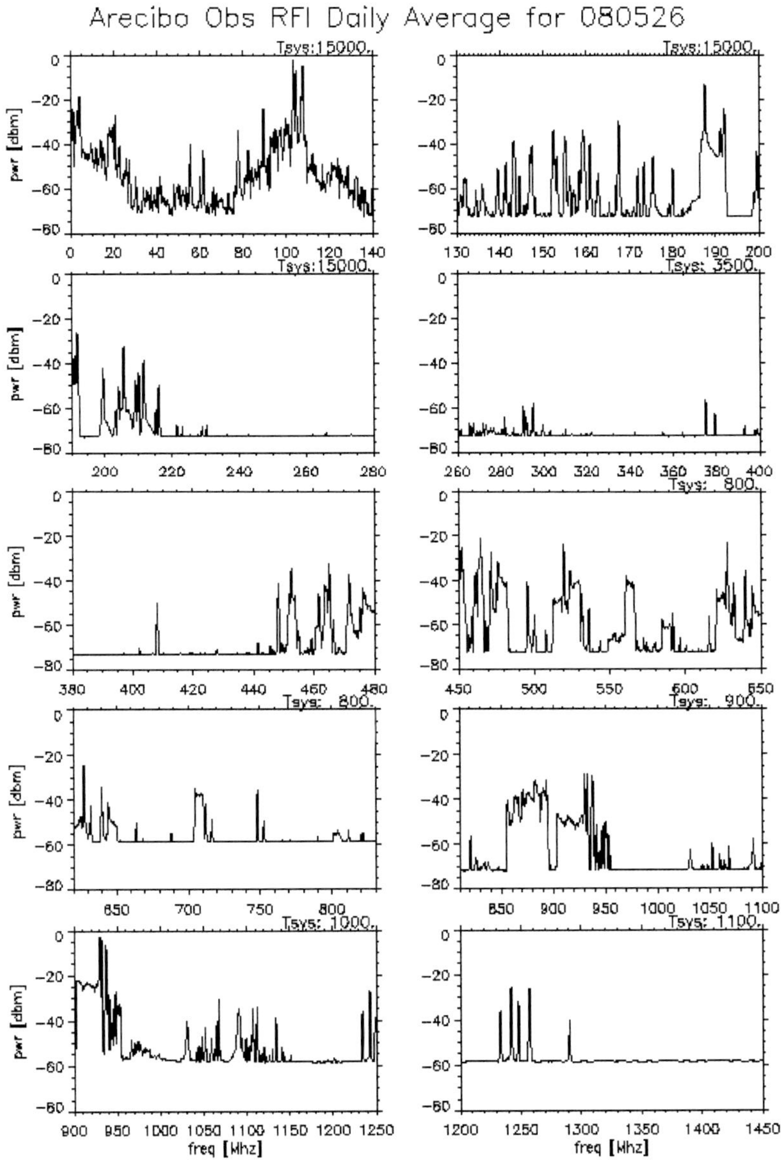

FIGURE 4.2 An example of radio frequency interference measurements made at the Arecibo radio observatory in Puerto Rico on May 26, 2008. The scan from a single location and a single instance in time is from a few megahertz to 1.45 GHz indicating the large number of commercial, government, and consumer uses of the spectrum. Detailed, real-time characterization of the spectrum uses provides an opportunity to prevent unauthorized uses of the spectrum from potentially causing catastrophic interference, as well as the capacity for opportunistically using unused spectrum for enhancing measurements. The Arecibo Observatory is part of the National Astronomy and Ionosphere Center, which is operated by Cornell University under a cooperative agreement with the National Science Foundation.

the passive and active user communities have a common interest in improving the ability to measure the utilization of the radio frequency spectrum.

A time resolution of 1 ms would be able to resolve and potentially classify transmit bursts in most mobile radio communications systems using time-division multiplexing (TDM) duplexing or channelization. Such systems use bursts/packets/frames in lengths of 10 ms to 40 ms due to a trade-off between accuracy in tracking propagation channels and throughput efficiency (payload/header ratio).[11]

However, a 1 millisecond time resolution would not resolve radar pulses, as these pulses are typically in the 2 to 400 microseconds range.[12] Furthermore, if pulses cannot be resolved, it would be more difficult to positively identify the source as radar, as opposed to intermodulation from other things that just happen to be emitting into that frequency. If the pulses are resolved, however, it becomes very easy to identify the source and also to determine whether the sources are "splattering," "jabbering," or exhibiting other illicit behaviors. For the purposes of the RAS and EESS, and conceivably for many other applications as well, the activity of these radar pulses is of great interest. Even the multipath from these radars can be problematic to sensitive systems.[13] To resolve these radar pulses, a time resolution on the order of 1 microsecond would be needed, which would not be technologically difficult to achieve.

The bandwidth resolution needed for such a spectrum survey could reasonably be 1 kHz; 1 kHz is roughly an order of magnitude less than the minimum standard bandwidth for any communications system above 30 MHz. A bandwidth resolution of 1 kHz would also resolve most communications below 30 MHz. Also, since much RFI comes in the form of unmodulated carriers (e.g., spurious products from transmitters, stuck microphones, etc.), the bandwidth is lower-bounded only by transmitter phase noise and so can be very narrow. A higher bandwidth resolution of, say, 10 kHz, would be too coarse, since most Land Mobile Radio Systems (two-way radios below 1 GHz) are migrating to 6.25 kHz channelization over the next decade.

Spatial resolution is the hardest parameter of the space to "saturate" with a monitoring system. Modern cellular systems use cell sizes ranging from building size to tens of kilometers. Satellite- and HAP-based cell systems can have cells hundreds of kilometers in extent. For terrestrial systems, this is highly frequency-,

[11]T.S. Rappaport, *Wireless Communications: Principles and Practice*, 2nd Ed., Prentice Hall, 2002.

[12]Frank H. Sanders, "Detection and Measurement of Radar Signals: A Tutorial," 7th Annual International Symposium on Advanced Radio Technologies, March 1, 2005; S.W. Ellingson and G.A. Hampson, "Mitigation of Radar Interference in *L*-Band Radio Astronomy," *Astrophysical Journal Supplement Series*, 147: 167-176 (July 2003); G. Miaris, T. Kaifas, Z. Zaharis, D. Babas, E. Vafiadis, T. Samaras, and J.N. Sahalos, "Design of Radiation-Emission Measurements of an Air-Traffic Surveillance Radar," *IEEE Antennas and Propagation Magazine*, Vol. 45, No. 4 (August 2003).

[13]S.W. Ellingson and G.A. Hampson, "Mitigation of Radar Interference in L-Band Radio Astronomy," *Astrophysical Journal Supplement Series*, 147: 167-176 (July 2003).

terrain-, and protocol-dependent: different systems have different transmitter densities and different typical transmitting powers. Any justifiable angular resolution requirement would be frequency-dependent, such that the survey would achieve lower resolution at lower frequencies and higher resolution at higher frequencies. This relationship has to do with the nature of multipath scattering versus frequency as well as with fundamental limitations in angular resolution—resolution improves with increasing aperture in wavelengths. The ability to locate emitters with sufficient accuracy to facilitate the identification of sources would be the goal of the survey, and given the dependencies mentioned above, the necessary spatial resolution would depend on the frequency and on what can be afforded.

Finding: Greater efforts to collect and analyze radio emission data are needed to support the enforcement of existing allocations and to support the discussion and planning of spectrum use.

Finding: Better utilization of the spectrum and reduced RFI for scientific as well as commercial applications are possible with better knowledge of actual spectrum usage. Progress toward these goals would be made by gathering more information through improved and continuous spectral monitoring. This would be beneficial to both the commercial and the scientific communities.

4.2 MAJOR DRIVERS OF SPECTRUM USE

Current measurements of spectral utilization and its impact on passive systems may not be indicative of future spectral use. The drivers for additional spectral bands for intensive use, the allocation of additional bands, and the development of "smart" flexible radio technology will have a profound impact on future use. The following assessment for the time period 2008-2015 is based on well-established drivers, currently allocated spectral bands, and technology that is under development. This assessment has a high-to-moderate level of confidence. That said, the impact from regulatory changes can be profound—for example, increases in power levels or emission levels permitted outside the primary transmission band could create an RF environment much less useful for passive systems.

Assessment of Trends in Spectrum Use for 2008-2015

The current trend toward more intensive use of the RF spectrum will continue unabated for both commercial and government uses. Within the United States, the continued desire for higher levels of access to the Internet (Figure 4.3), coupled with the increased desire for mobility, will incite the development of new commercial systems. Technology is also a major driver for more intensive use. New mobile

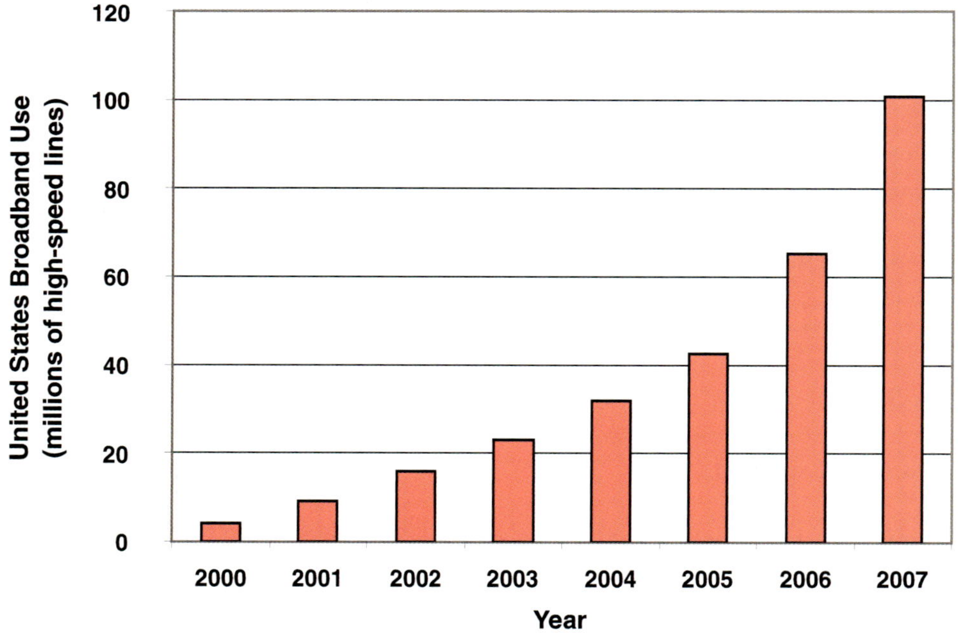

FIGURE 4.3 Broadband services usage in the United States from 2000 through 2007, as indicated by the number of high-speed lines (or wireless channels) over 200 kbps in at least one direction. SOURCE: Federal Communications Commission, *High-Speed Services for Internet Access Report,* March 2008, p. 6, available at http://hraunfoss.fcc.gov/edocs_public/attachmatch/DOC-280906A1.pdf.

devices can integrate the use of multiple modes and bands within a single handset, and they use ever-wider-band RF components that allow for higher data rates. Such devices will drive the continued desire for more spectrum for commercial activities. The combination of mobility and integration has a strongly deleterious effect with respect to passive systems, because it will create a more pervasive use of the spectrum not only in spectral extent but in geographic extent as well.

The greater impact on the EESS community from more intensive use of the spectrum will come from what is occurring in developing nations. Fixed-line infrastructure including copper and fiber is available in highly developed countries except in low-population-density regions. This infrastructure is not available in developing nations, but unused radio spectrum is readily available. Therefore, most commercial deployments, including backhaul, are made entirely out of wireless systems. Economic development in these nations will produce a much higher reliance on wireless systems and will see a much higher growth rate in the use of wireless transmission systems. One example is that in 2006, China had more new cellular subscribers than the total number of U.S. subscribers (see Figure 4.4). The

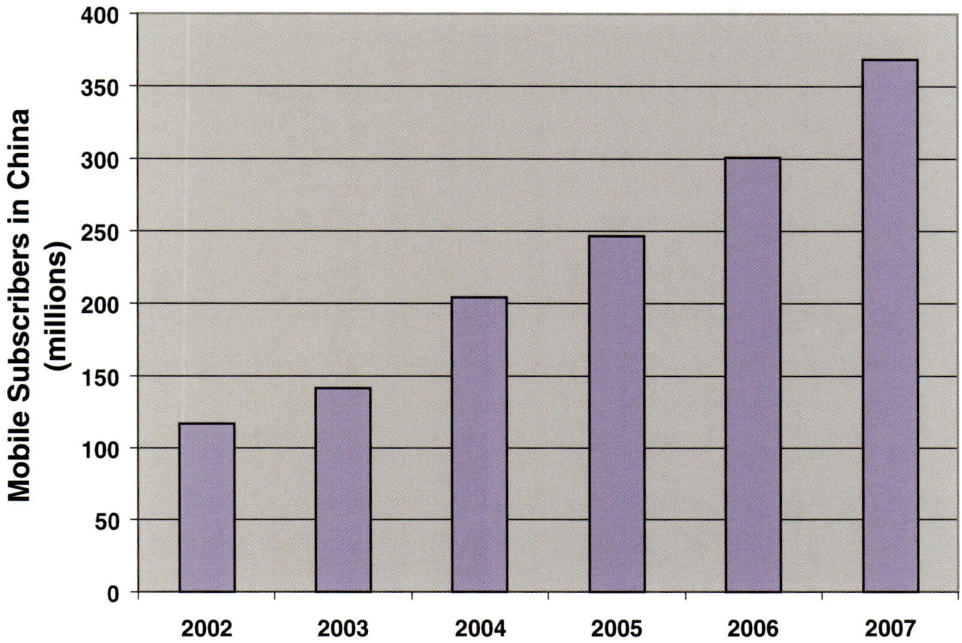

FIGURE 4.4 The number of cellular subscribers in China grew by more than 250 million between 2002 and 2007. SOURCE: *China Mobile Ltd. Annual Reports.*

ITU has indicated that the number of mobile cellular users worldwide at the end of 2007 was in excess of 3.3 billion.[14]

Lastly, the mechanism by which spectrum licenses are obtained has a large impact on what is optimized for system deployment. For example, in 2006 the FCC auctioned 90 MHz of spectrum for Advanced Wireless Services (AWS; also known as Third Generation cellular). The auction netted the U.S. Treasury $13.7 billion, which is equivalent to $0.50 MHz-pop.[15] This cost is usually called the opportunity cost for the spectrum. High opportunity costs motivate the licensee to use the spectrum quite efficiently to leverage the already-"sunk costs." This is why cellular operators are very conscious of their spectral efficiency and thus exploit spectral

[14]"Global Mobile Phone Users Top 3.3 Billion by End-2007—Study," *Cellular News,* May 24, 2008, available at http://www.cellular-news.com/story/31352.php; accessed January 14, 2010.

[15]The figure "$0.50 per MHz-pop" indicates that the licenses, if amortized across the entire U.S. population, would be worth $0.50 for every megahertz of bandwidth (it would be somewhat higher if one takes into account the effective population of subscribers). However, with 200 million subscribers and a maximum market penetration of approximately 30 percent, it would yield a value of $2.53 per megahertz per active subscriber.

reuse techniques by using the same frequency bands at each tower. However, access to spectrum without any opportunity costs tends to motivate the use very differently. For example, the spectrum for public-safety users is provided as a direct grant. These users usually deploy high-site, high-power transmitters to reduce the cost of the infrastructure. The public-safety services are less efficient spectrum users than are the cellular telephone services, since the technology of the former lacks any spectral reuse.

Third-Generation and Fourth-Generation Systems

The 2008-2015 period will see the deployment of new cellular-based services in bands that were allocated in the 2002-2007 period: the 700 MHz band, the AWS bands, and the Broadband Radio Service and Educational Broadband Service (BRS/EBS) bands. These bands were reallocated by the FCC in recognition of the rapid growth in demand made by mobile data services.

700 MHz

The digital TV transition that took place in June 2009 recaptured the spectrum now allocated as channels 52-69 (698-806 MHz). The propagation characteristics of these bands, as well as the lack of incumbents within the band, make this piece of the spectrum highly prized. The band has allocations for Public Safety (763-775 MHz and 793-805 MHz), moderate-power[16] cellular operations (746-763 MHz and 776-793 MHz), and high-power[17] operations (698-743 MHz). Much of the band was yet to be licensed as of December 2008. In March 2008, an auction took place in which $19.6 billion was bid for the licenses within this band. However, licenses that do not meet the build-out requirements to provide coverage to a required percentage of the U.S. population will be remanded to the FCC, leading to expectations that these bands will be intensively used.

Advanced Wireless Services 1, 2, and 3

The Advanced Wireless Services (AWS) bands include 1710-1755 MHz, 1915-1920 MHz, 1995-2000 MHz, 2020-2025 MHz, and 2110-2180 MHz. In September 2006, the FCC auctioned the 1710-1755 MHz band paired with the 2110-2155 MHz band and denoted it the AWS-1 block. The AWS band is generally called the "3-G" (Third Generation) band and is denoted for mobile voice and data services. The build-out requirements for AWS-1 are not as cumbersome as those

[16]Up to 2,000 W/MHz in rural deployments.
[17]Up to 50 kW for cellular broadcast deployments.

for the 700 MHz band. Additional auctions have been proposed for the AWS-2 and AWS-3 bands.

2.5 GHz: Broadband Radio Service and Educational Broadband Service

The Multipoint Distribution System (MDS); Multi-channel, Multipoint Distribution System (MMDS); and Instructional Television Fixed Service (ITFS) bands that formerly occupied the 2.1 GHz spectrum were reallocated to the 2495-2690 MHz band. The transition also included a name change to the BRS and EBS. Wireless Internet access is the primary use of these bands, which have rules that permit technical flexibility to deploy any technology that meets the emission rules. However, WiMAX (also known as IEEE 802.16) technology is generally the technology that is deployed.

Unlicensed Uses of the Radio Frequency Spectrum

There has been a phenomenal proliferation of unlicensed devices over the past decade, to the great benefit of society.[18] Over the past 5 years there have also been significant additions to the types of unlicensed devices and spectral bands available for unlicensed uses. The ultrawideband types of unlicensed devices, the expansion of unlicensed use in the Unlicensed National Information Infrastructure (U-NII) band, and the allocation for 70, 80, and 90 GHz bands are noteworthy.

Unlicensed devices operate at a low enough power to be deemed not harmful to the primary licensee in the band. Unlicensed devices can also operate on a co-primary basis in certain bands. The 900 MHz, 2.4 GHz, and 5.8 GHz bands are three such bands. The 900 MHz band has been popularly used by baby-monitoring device and wireless phone manufacturers. The 2.4 GHz band is the most popular one and is heavily used by wireless local-area networks (also known as Wi-Fi), cordless telephones, security systems, and personal-area networks (also known as Bluetooth) manufacturers. Indeed, unlicensed devices have many societal and commercial benefits.

Ultrawideband

Ultrawideband (UWB) is a technology for transmitting information using a large bandwidth (>500 MHz), which can cross many spectrum-allocation boundaries. UWB was originally accepted as pulse radio, but the FCC and ITU-Radio (ITU-R) now define UWB in terms of a transmission from an antenna for which

[18] *Unlicensed devices* are those that are allowed to operate without a specific license for a particular spectral band. They are also sometimes called license-free devices.

the emitted signal bandwidth exceeds the lesser of 500 MHz or 20 percent of the center frequency. The FCC authorizes the unlicensed use of UWB in the 3.1-10.6 GHz band, and the FCC power spectral density emission limit for UWB emitters operating in the UWB band is –41.3 dBm/MHz. These emission limits are consistent with those granted by the FCC for personal computer emissions and intentional emissions for unlicensed devices.

Unlicensed National Information Infrastructure at 5 GHz

The FCC established a schedule for new unlicensed devices that are dynamic frequency selection (DFS)-compliant in the 5.25-5.35 GHz (UNII-2) band and in a new spectral region between 5.470 and 5.725 GHz (UNII-3). [19] The new DFS rule is required in order to allow the coexistence of unlicensed devices with existing military and weather radar systems in the 5 GHz band. The new FCC rule requires that unlicensed devices must comply with DFS to prevent the devices from interfering with incumbent military and weather radar systems. The DFS system must continuously monitor the selected frequency channel during use and if a radar signal is detected, the DFS system must stop and jump to another available channel that has gone through the same selection process.

Millimeter Wave: 70, 80, 90 GHz

In October 2003, the FCC opened 13 GHz of previously unused spectrum at 71-76 GHz, 81-86 GHz, and 92-95 GHz, for high-density fixed wireless.[20] Although not explicitly used for unlicensed devices, millimeter-wave communications have many of the characteristics of unlicensed use: many of the requirements in obtaining licenses are minor and would essentially allow a great deal of proliferation of those devices. The FCC will issue an unlimited number of non-exclusive nationwide licenses to non-federal government entities for the 13 GHz of spectrum allocated for commercial use. These licenses will serve as a prerequisite for registering individual point-to-point links. The 71-95 GHz bands are allocated on a shared basis with federal government users. Therefore, a licensee will not be authorized to operate a link under its non-exclusive nationwide license until the link is coordinated with the NTIA with respect to federal government operations and is registered as an approved link with a third-party database manager.

[19]See, 47 C.F.R. Sec. 15.407. See also, *In the Matter of Revision of Parts 2 and 15 of the Commission's Rules to Permit Unlicensed National Information Infrastructure (U-NII) Devices in the 5 GHz Band,* Memorandum Opinion and Order, 21 FCC Rcd 7672 (2006).

[20]*In the Matter of Allocations and Service Rules for the 71-76 GHz, 81-86 GHz and 92-95 GHz Bands; Loea Communications Corporation Petition for Rulemaking,* Report and Order, 18 FCC Rcd 23318 (2003).

Regulatory Changes That Impact Use

As previously described, the regulatory agencies (NTIA, FCC) determine the appropriate technical parameters for operation in each spectral band. These rules are in constant flux in order to keep current with technological changes and national needs. Since 2005, the regulatory agencies have been moving away from explicit emissions parameters. This change is in recognition of the fact that it is not power but power spectral density and power flux density that better represent proper operating parameters:

- *Power spectral density:* In 2007, the FCC changed emission rules for AWS and 700 MHz from 1640 W equivalent isotropically radiated power (EIRP) to 1640 W/MHz EIRP. This change was in response to the penalty to broadband systems.[21] The transmitter power of broadband systems was regulated regardless of bandwidth, so these systems would be afforded a lower transmitter power spectral density than that for narrowband systems.
- *Power flux density:* The FCC has begun to use power flux density (e.g., Wm^{-2}) as the key emission parameter. This is in response to the previous lack of incentive for using elevation beam shaping to control the interference at ground level and to allow higher power emission limits for more cost-effective commercial deployments. The technical rules for the 700 MHz spectral band allow a transmission power up to 50 kW in portions of the band, but the power flux density must be less than 3,000 microwatts per square meter on the ground.[22]

These rule changes represent an opportunity to use more sophisticated interference metrics for interference control. The recent changes increased sophistication with both spectral and spatial characterization. It may also be possible to extend regulations to include temporal characterizations that will be useful for developing new interference mitigation techniques.

The regulatory environment has investigated but has yet to address three additional means for interference mitigation:

- *Interference metrics:* Metrics have been investigated to clarify what is considered to be harmful interference. Currently the regulators primarily quantify transmitter characteristics in lieu of explicit interference control. One proposal, called Interference Temperature, was closely related to noise

[21] Using a power-only metric versus a power spectral density metric essentially allowed smaller bands (e.g., 5 MHz wide) to have twice the power of a 10-MHz-wide band.

[22] 47 C.F.R., Vol. 2, Sec. 27.55 (b).

temperature, which is used extensively by the RAS community.[23] An engineering-based metric would provide clarity for system developers and for policy makers in determining the relative value of systems. The metric could be different for different bands and applications.

- *Regulatory enforcement:* Current means of enforcement are primarily by licensee self-enforcement or by the FCC's use of a limited number of mobile interference-monitoring laboratories (seven in the United States). The proliferation of mobile wireless transmitters within consumer, commercial, and government systems requires new monitoring and enforcement technologies.
- *Inclusion of passive systems in regulatory databases:* Current FCC databases include all transmission equipment for site-specific licenses but do not include passive systems such as those used in the EESS and RAS. These databases are used by licensees to determine the potential for interference between systems and to communicate with other licensees on a case-by-case basis. Since the passive systems are not included in the FCC databases, they are not considered in these discussions. Knowledge of the location and operational characteristics of the passive systems would be very useful to licensees for determining impact and possible mutual interference mitigation techniques.

The new techniques for interference control have been investigated by regulators but have not been acted on. These techniques include interference metrics (e.g., interference temperature), improving enforcement technology to provide new

[23](1) Federal Communications Commission (FCC), "Spectrum Policy Task Force," 2002; FCC, Docket 03-237, Notice of Inquiry and Notice of Proposed Rulemaking on the Establishment of an Interference Temperature Metric to Quantify and Manage Interference and to Expand Available Unlicensed Operation in Certain Fixed, Mobile, and Satellite Frequency Bands; (2) P.J. Kolodzy, "Interference Temperature: A Metric for Dynamic Spectrum Utilization," *Int. J. Netw. Manag.*, 16(2): 103-113 (2006); (3) T.C. Clancy, "Achievable Capacity Under the Interference Temperature Model," *INFOCOM 2007 26th IEEE International Conference on Computer Communications*, May 6-12, 2007; (4) Joe Bater, Hwee-Pink Tan, Kenneth N. Brown, and Linda Doyle, "Maximising Access to a Spectrum Commons Using Interference Temperature Constraints," *Cognitive Radio Oriented Wireless Networks and Communications 2007, 2nd International Conference*, pp. 441-447, Aug. 1-3, 2007; (5) Yiping Xing, C.N. Mathur, M.A. Haleem, R. Chandramouli, and K.P. Subbalakshmi, "Priority Based Dynamic Spectrum Access with QoS and Interference Temperature Constraints," *IEEE International Conference on Communications*, 10: 4420-4425 (June 2006); (6) J.A. Stine, "Spectrum Management: The Killer Application of Ad Hoc and Mesh Networking," *First IEEE International Symposium on New Frontiers in Dynamic Spectrum Access Networks*, pp.184-193 (November 8-11, 2005); (7) A.E.E. Rogers, J.E. Salah, D.L. Smythe, P. Pratap, J.C. Carter, and M. Derome, "Interference Temperature Measurements from 70 to 1500 MHz in Suburban and Rural Environments of the Northeast," *First IEEE International Symposium on New Frontiers in Dynamic Spectrum Access Networks*, pp. 119-123, (November 8-11, 2005).

tools for the regulators to ensure compliance with emission rules (e.g., commercial devices used for enforcement measurements, additional mobile measurement systems, etc), and the inclusion of passive systems in regulators' databases (e.g., the FCC's Universal Licensing System).

Finding: Current regulatory structure and support infrastructure (such as databases, etc.) are transmitter-centric. Methodologies to incorporate passive systems need to be developed.

Technology Changes That Impact Use

Software-Defined Radios and Cognitive Radios

The development of wideband power amplifiers, synthesizers, and analog/digital (A/D) converters is providing for a new class of communication radios defined by software: the software-defined radio (SDR) and cognitive radio (CR). Although at the early stages of development, this new class of radio ushers in new possibilities as well as possible pitfalls for technology policy. The flexibility provided by the CR class of radio allows for more dynamics within radio operations. The technology also makes possible dynamic collaboration between CR and science users. The same flexibility poses challenges for certification and the associated liability through potential misuse.

SDR provides software control of a variety of modulation techniques, wideband and narrowband operation, communications security functions (such as hopping), and waveform requirements. In essence, components can be under digital control and thus defined by software. The advantage of an SDR is that a single system can operate under multiple configurations, providing interoperability, bridging, and tailoring of the waveforms to meet the localized requirements. SDR technology and systems have been developed for the military. The digital modular radio (DMR) system was one of the first SDR systems. Recently the Defense Advanced Research Projects Agency developed the Small Unit Operations Situational Awareness Systems, a portable SDR operating from 20 MHz to 2.5 GHz. The success of these programs has led to the Joint Tactical Radio System initiative to develop and procure SDR systems throughout the U.S. military.

SDRs exhibit software control over a variety of modulation techniques and waveforms. Software radios can specifically implement the signal processing in software and use digital-to-analog converters to translate from the digital domain to the RF domain. This additional capability essentially has the radio being constructed with an RF front end, a down-converter to an intermediate frequency or baseband, an A/D converter, and a processor. The processing capacity limits the complexity of the waveforms that can be accommodated.

A cognitive radio adds both a sensing and an adaptation element to the software defined radios and software radios. Four new capabilities embodied in cognitive radios will help enable dynamic use of the spectrum: flexibility, agility, RF sensing, and networking.[24]

- *Flexibility* is the ability to change the waveform and the configuration of a device;
- *Agility* is the ability to change the spectral band in which a device will operate;
- *Sensing* is the ability to observe the state of the system, which includes both the radio unit and more importantly, the RF environment; and
- *Networking* is the ability to communicate between multiple nodes and thus facilitate the combining of the sensing and control capacity of those nodes.

These new technologies and radio classes, albeit in their nascent stages of development, are providing many new tools to the system developer while allowing for more intensive use of the spectrum. However, an important characteristic of each of these technologies is the ability to change configuration to meet new requirements. This capacity to react to system dynamics will forever change how new uses of the RF spectrum are addressed.

Microelectromechanical Systems Filters

Microelectromechanical systems (MEMS) involve the integration of mechanical elements, sensors, actuators, and electronics on a common silicon substrate through microfabrication technology.

Recent technology developments have demonstrated the potential to create high-performance MEMS acoustic resonators. Other MEMS filter technologies under development include tunable resonators and multi-pole tunable filters, the integration of filter arrays with switch arrays to create large filter banks, and the integration of filter banks with active RF devices to form complete RF front ends. MEMS-based mechanical resonators and filters have shown promising characteristics in achieving high-Q values and good stability.

Digital Modulation and High-Efficiency Modulation Schemes

The signals received by radio astronomy and passive Earth remote sensing are random. The signals follow Gaussian statistics within a given bandwidth and

[24]International Telecommunication Union, "Techniques for Mitigation of Radio Frequency Interference in Radio Astronomy," Document 7D/142-E, January 23, 2007.

usually have a flat, featureless spectrum. Until recently, these characteristics distinguished natural signals from human-made signals, and this fact could be used as a tool to distinguish between the wanted natural signals and interference from human-made emissions.

The move to digital modulation, and in particular to high-efficiency digital modulation schemes, has the effect of making the human-made signals indistinguishable from the wanted natural signals. The statistical fluctuations with time, and the flat, featureless power spectrum displayed by such signals are nearly the same as the characteristics of the wanted natural signals that are the subject of the research. This situation has two detrimental effects:

- One of the distinguishing characteristics of human-made interference, the presence of distinct and discrete carrier waves, has disappeared. The mitigation of interference is now made more difficult.
- The higher spectral efficiency of modern modulation schemes is generally a very good thing. However, passive use-based research has often taken advantage of the inefficient modulation schemes by, for example, making use of small gaps between carriers in spectrum used by the active services. The greater efficiency of current spectrum use by the active services as enabled by higher-order digital modulations has unfortunately made passive use difficult or impossible.

Finding: The emergence of practices for the dynamic use of the spectrum will result in more devices with greater variability in active spectrum usage, and the EESS and RAS communities could be impacted with more unintentional radio interfering devices.

Finding: New cooperative spectrum management techniques that could be beneficial for enhanced interference management and increased spectral utilization have been investigated by regulators but have not been implemented.

Summary

The RAS and EESS have current spectrum allocations between 9 kHz and 3 GHz of approximately 62.12 MHz on a primary basis and 122.5 MHz on a secondary basis. However, the spectrum that the radio astronomy and Earth exploration radio science community currently use far exceeds these allocations. The use of new techniques requiring other spectral bands or additional spectrum is driven by the scientific requirements and made possible by continual technical developments, as discussed in Chapters 2 and 3. Since the RAS and EESS are passive services, they operate outside their allocations on a non-interference basis when the RFI is weak enough.

The strong increase in the number of devices and systems deployed for consumer, commercial, and industrial uses has fueled an interest in learning how the radio spectrum is actually used. While the various efforts of the active and passive user communities have been useful in confirming the sparse time-frequency utilization of the spectrum, most existing studies are of limited usefulness because of their limited sensitivity, time resolution, and frequency resolution. Useful spectral monitoring for RAS applications requires continuous time-frequency resolution on the order of 1 kHz and 1 microsecond, with sensitivity sufficient to detect signals approaching the levels already known to be potentially harmful to radio astronomy as determined in ITU-R Recommendation RA.769. For air- and spaceborne EESS applications, an adequate spatial resolution would be critical, as discussed in §4.1.

The major drivers for more intensive spectral use will come from newly deployed 3G and 4G cellular systems, new technologies for unlicensed devices (Wi-Fi, etc.), changes to the regulations, and the availability of advanced technologies such as cognitive radios. It should be expected that the use of the 700 MHz, 1710-1755 MHz, 2110-2180 MHz, and 2495-2690 MHz spectral bands will be greatly increased due to 3G and 4G cellular deployments. Unlicensed devices will continue to proliferate in the 5 GHz, 70-90 GHz, and 3.1-10.6 GHz (ultrawideband) spectral bands. The onset of agile, frequency-hopping radio technology will create challenges: the prediction of open spectral bands will become more difficult. Regulatory agencies determine the appropriate technical parameters for operation in each spectral band, but these rules are in constant flux. The agencies need to enhance their role with respect to the passive services to include interference metrics, an extension of enforcement technology, and the inclusion of passive systems in their databases (e.g., the FCC's Universal Licensing System).

4.3 UNILATERAL MITIGATION TECHNIQUES

A variety of techniques have been developed to reduce the impact of RFI on EESS and RAS observations. This section presents a review of unilateral methods. These methods apply to situations in which the EESS or RAS operator has no ability to influence the behavior of the sources producing the interference. This is the most common situation, but at present the performance achieved by the majority of the unilateral methods has been documented only anecdotally and so remains to be completely quantified.[25] Following the review of specific unilateral mitigation

[25]International Telecommunication Union, "Techniques for Mitigation of Radio Frequency Interference in Radio Astronomy," Document 7D/142-E, January 23, 2007; A.J. Boonstra, "Radio Frequency Interference Mitigation in Radio Astronomy," PhD Thesis, Delft University of Technology, Dept. EEMCS, June 2005, ISBN 90-805434-3-8.

technologies is a summary (in Table 4.2) of the successes that have been achieved as well as the inherent limitations of particular approaches.

Before proceeding further, it is important to distinguish among the various RFI environments and observational situations encountered by EESS and RAS systems. Both services are subject to variations in the RFI environment caused by changes in RFI source behaviors as well as by changes in the position of these sources (including the orbital motions of spaceborne RFI sources). However, EESS use is both ground- and space-based and needs access to areas on a global basis. Space-based radiometers operate in a conically scanning configuration in which the portion of Earth's surface observed (and associated RFI sources) varies within the timescale of a few milliseconds—the duty factor for a given area observed. In contrast, RAS antennas typically operate at a fixed position on the ground and in a viewing configuration that is stable over timescales of several minutes to many hours in order to receive extremely weak signals.

The "downlooking" nature of EESS measurements makes the probability of ground-based RFI sources being within the main beam of the antenna a common occurrence, while the primary concern for main-beam corruption in RAS applications is spaceborne sources, which are encountered less frequently. The "uplooking" RAS systems are inherently more sensitive to RFI, owing both to the cold background sky (relative to the hot ground emission in EESS sensing) and to the typically long integration times. Main-beam corruption of EESS measurements, due to a reflection of space-based sources, has also been observed.[26] Both systems are subject to the influence of ground- and space-based sources received through antenna sidelobes. RAS measurements are often interferometric, using antennas separated over distances that span continents and are millions of wavelengths in extent, whereas EESS radiometers typically use only a single antenna, and even interferometric systems in EESS are limited to maximum spatial separations on the order of meters (up to a few thousand wavelengths) due to the limited spatial extent available on a space- or airborne platform. Finally, systems operating in space are subject to restrictions on the output data rate not usually encountered by systems operating on the ground.

Technologies for Unilateral Mitigation

Unilateral mitigation methods involve two primary components: detection and suppression. The former involves the determination that a particular observation contains RFI, whereas the latter attempts to utilize information from the detection

[26]L. Li, P.W. Gaiser, M.H. Bettenhausen, and W. Johnston, "WindSat Radio-Frequency Interference Signature and Its Identification over Land and Ocean," *IEEE Transactions on Geoscience and Remote Sensing*, 44(3): 530-539 (March 2006).

stage in order to correct observed data by removing RFI contributions. The study of techniques both for detecting and suppressing RFI has been a topic of heightened interest in recent years, spurred by technological advances that enable real-time signal-processing approaches. Strong RFI sources that produce excessively high observed powers tend to be easily distinguished from astronomical or geophysical signals. In those cases, the detection problem is relatively straightforward. The detection of weak RFI sources that produce power levels comparable with those associated with geophysical or astronomical variability is much more difficult. This is problematic because, if not suppressed, weak RFI can still introduce significant errors into the scientific conclusions that are drawn from the measurements, often without warning.

It is very important to note that unilateral mitigation techniques do not and cannot solve the RFI problem. However, they can serve several important purposes. They can provide a means of limiting the introduction of corrupted observations into the scientific user community. They can provide significant relief from the labor-intensive task of manually identifying the effects of RFI on science data products. And they can permit some limited scientific uses of very noisy, interference-laden portions of the spectrum that would otherwise not be possible. But in most scientific applications, RFI mitigation techniques cannot actually remove the interference from those portions of the spectrum where it is present. Thus it can be expected that as RFI becomes more prevalent, the need for interference mitigation in order to do useful science will increase, but at the same time the efficacy of mitigation efforts will decrease and the quality of the resulting science will suffer accordingly.

Detection Techniques

Various RFI detection techniques are available, with each typically oriented toward a particular class of RFI sources. A classical detection algorithm takes observed data as input and provides a binary "yes/no" output as to whether RFI is present. The input data can range from Nyquist sampled received voltages to final output powers integrated over millisecond or longer timescales. Detection algorithms involve a trade-off between the probability of detecting a specified type of RFI and the probability of obtaining a "false alarm" in which RFI-free data are erroneously classified as corrupted. An excessive false-alarm rate can lead to a reduction of measurement sensitivity; this can be addressed by modifying the detector, but typically at the cost of a reduced probability of detecting true RFI. Specific classes of detection techniques are described below.

Matched Filtering for Known Sources

The problem of detecting a specified signal in additive Gaussian noise (i.e., thermal noise) is well defined and has a long history of investigation in the signal-processing literature. The best detection performance that can be achieved is obtained through a "matched filter" approach, in which the detector is a filter designed to have a frequency response conjugate to that of the signal of interest. This approach is the standard method for use in communications applications, but it has limited applicability in radio science applications because the RFI sources encountered are not always known a priori.

Tests for Gaussianity

Instead of attempting to detect particular RFI signals, it is possible to detect whether observed voltage appears to have originated from a thermal noise (i.e., Gaussian) field distribution. This is a classical problem in statistics, and numerous tests are available, with no particular test having been shown clearly superior for EESS and RAS systems. One example of this technique is a "kurtosis test," which compares the normalized fourth moment of the observed voltage to that expected for thermal noise fields. This method has been shown to provide good performance in tests using ground-based EESS instruments, and the expected detection performance for pulsed-sinusoidal RFI interference has been analyzed in detail.[27] However, the performance of this approach for other RFI source types remains to be quantified. An analog implementation of the kurtosis detector has also been described,[28] as has a kurtosis method combined with multiple frequency channels for RAS applications.[29] It should be expected that the modulation-insensitive nature of Gaussianity tests will result in a detection performance that is at best equal to that of detection algorithms designed for a priori known RFI source types. It is often the case that the nature of the RFI is not known beforehand. In these cases, tests for Gaussianity have been found to be a robust RFI detector.

Pulse Detectors

This is perhaps the oldest and best-known strategy for the detection of RFI, whether at timescales of the Nyquist sampled receiver bandwidth or at that of the final output data product. Typically, an "acceptable" range for the received data amplitude as a function of time is defined, and points outside this range are deemed

[27]C.S. Ruf, S.M. Gross, S. Misra, "RFI detection and mitigation for microwave radiometry with an agile digital detector," *IEEE Transactions on Geoscience and Remote Sensing*, 44(3): 694-706, (March 2006); and R.D. De Roo, S. Misra, C.S. Ruf, "Sensitivity of the Kurtosis Statistic as a Detector of Pulsed Sinusoidal RFI," *IEEE Transactions on Geoscience and Remote Sensing*, 45(7, Part 1): 1938-1946 (July 2007).

[28]Jeffrey Piepmeier, Priscilla Mohammed, and Joseph Knuble, "A Double Detector for RFI Mitigation in Microwave Radiometers," *IEEE Transactions on Geoscience and Remote Sensing*, 46(2): 458-465 (2007).

[29]G. Nita, D.E. Gary, Z. Liu, G. Hurford, and S.M. White, "Radio Frequency Interference Excision Using Spectral Domain Statistics," *Publ. Astron. Soc. Pacific*, 119: 805-827 (2007).

as corrupted. The acceptable range can be defined in either an absolute sense (i.e., fixed thresholds) or a relative sense (i.e., as a number of standard deviations around a local mean value). Pulse detectors operating at high sample rates are well suited for the detection of low-duty-cycle radar emissions; such sources typically transmit pulsed fixed-frequency or chirped sinusoidal waveforms with pulse lengths of 2-400 microseconds, 1-27 ms between transmitted pulses, and bandwidths on the order of 1 MHz. For low-duty-cycle pulsed interferers, the sensitivity of the detector depends on the relationship between the timescales of the pulsed RFI and the input data (i.e., the sample rate). Good detection performance can be achieved in cases where the individual pulses are resolved by the input data sample rate. A number of pulse detection techniques have been proposed and developed to various degrees. Pulse detection is, however, not at all appropriate for RFI sources that are continuous in nature.

Passive receiver blanking has been attempted using a special receiver and perhaps also a special antenna directed at the source of interference. This might, for example, mitigate interference from a pulsed radar transmitter. When the pulses from the radar are received, the radio astronomy receiver electronically triggers a data masking or data flagging process. The limitations of this method are primarily that it requires excellent sensitivity on the unwanted transmission, and secondly that it may be hard to accommodate transmissions arriving with different delays from different directions, such as multipath propagation with multiple reflections from surrounding mountains.[30] In this case active receiver blanking using a beacon transmission on some carefully chosen frequency could be used at the radio astronomy site to blank the radio astronomy receiver. Note that the above scheme would not be effective for EESS due to the need for full global coverage.

Narrowband Source Detectors

Narrowband detectors are analogous to pulse detection methods but are better suited to signals that can be resolved in frequency—that is, searching for "outliers" among data in multiple frequency channels. Such approaches are designed to detect interference that is localized in frequency (i.e., narrowband), and performance is improved by matching the frequency resolution of the radiometer channels to that of expected RFI sources. Narrowband detectors record data in multiple frequency channels; these frequency channels can be achieved either by analog means (through use of a tuning receiver, a filter "bank," or a spectroscopy method to create a set of channels) or digitally (using either an internal Fourier transform or a set of digital filters). Example algorithms for detecting narrowband interference in

[30]J.R. Fisher, Q. Zhang, Y. Zheng, S.G. Wilson, and R.F. Bradley, "Mitigation of Pulsed Interference to Redshifted H I and OH Observations between 960 and 1215 MHz," *Astronomical Journal*, 129: 2940-2949 (June 2005).

a set of channel measurements have been described[31] and "spectral differencing" techniques have also been applied to detect RFI with the Advanced Microwave Scanning Radiometer-Earth (AMSR-E)[32] and WindSat radiometers[33] currently in orbit. If narrowband detection strategies are applied in postprocessing (i.e., not performed in real time by the radiometer), their use implies that the data rate of the radiometer is multiplied by the number of channels used; the resulting data rate can easily become prohibitive for space-based systems.

Narrowband detection algorithms are best for narrowband sources of large amplitude; however, the contribution of these sources to the total observed noise power can remain small due to the small ratio of the source bandwidth to that of the total radiometer channel. Performance is degraded for lower-amplitude RFI sources occupying bandwidths that are appreciable compared to the total radiometer channel bandwidth. RAS science requires the narrowband detection of spectral line emissions from atoms and molecules with Doppler shifts owing to the relative motion of Earth and object as well as Doppler spreading owing to kinematics internal to the object. Weak narrowband RFI can preclude, or at least corrupt, such measurements. Combinations of pulse and narrowband detection strategies are also possible.[34]

Polarization-Based Algorithms

The polarization properties of received radio waves also provide an opportunity for RFI detection. Geophysical and astronomical sources have polarization properties that, in many cases, can be predicted a priori to within a reasonable uncertainty. RFI sources that are highly polarized can create power differences among polarizations that can be recognized as anthropogenic. The success of such approaches depends on the extent to which the RFI source emissions appear polarized to the radiometer (which depends on RFI source properties, the orienta-

---

[31] A.J. Gasiewski, M. Klein, A. Yevgrafov, and V. Leuski, "Interference Mitigation in Passive Microwave Radiometry," *IEEE Geoscience and Remote Sensing Symposium, Conference Proceedings*, 3: 1682-1684 (2002); and B. Guner, J.T. Johnson, and N. Niamswaun, "Time and Frequency Blanking for Radio Frequency Interference Mitigation in Microwave Radiometry," *IEEE Transactions on Geoscience and Remote Sensing*, 45: 3672-3679 (2007).

[32] E.G. Njoku, P. Ashcroft, T.K. Chan, and Li Li, "Global Survey and Statistics of Radio-Frequency Interference in AMSR-E Land Observations," *IEEE Transactions on Geoscience and Remote Sensing*, 43(5): 938-947 (May 2005).

[33] Li Li, P.W. Gaiser, M.H. Bettenhausen, and W. Johnston, "WindSat Radio-Frequency Interference Signature and Its Identification over Land and Ocean," *IEEE Transactions on Geoscience and Remote Sensing*, 44(3): 530-539 (March 2006); and S.W. Ellingson and J.T. Johnson, "A Polarimetric Survey of Radio-Frequency Interference in C- and X-Bands in the Continental United States Using WindSat Radiometry," *IEEE Transactions on Geoscience and Remote Sensing*, 44(3): 540-548 (March 2006).

[34] S.M. Kay, *Fundamentals of Statistical Signal Processing: Volume II, Detection Theory*, Upper Saddle Creek, New Jersey: Prentice Hall (1998).

tion of both the source and receiver antennas, and the influence of multipath and other propagation effects), as well as the level of natural variations in polarization signatures for the medium observed. A polarization detection strategy has been used to detect interference in data generated by the EESS satellite AMSR-E,[35] but the strategy was found less sensitive to low-level interference than the spectral differencing method. An alternate polarization-based detection strategy based on the polarimetric channels in the radiometer of the EESS satellite WindSat was found to yield improved performance, because the small, geophysically expected polarized returns are readily exceeded by RFI sources.[36] However, such detection strategies remain dependent on antenna orientation and observation geometry issues, as well as on RFI source polarization properties. To date, polarization-based methods have generally received less attention than that given to other detection strategies.

Multiple-Antenna Algorithms

For instruments using multiple antennas, RFI detection algorithms can be developed on the basis of the relationships among the waves received at the antennas. A useful concept in searches for astronomical transients is anticoincidence,[37] in which the criterion for detection of astrophysical signals is that they appear in multiple, widely separated antennas, whereas terrestrial RFI should be relatively "local" and appear only in one or a subset of antennas. An inversion of this technique is used in an antenna with an array feed: the desired celestial signal is received in one of the many array feeds, but RFI is received in all of them. RAS synthesis imaging arrays such as the VLA and the Very Long Baseline Array have reduced sensitivity to RFI due to a lack of coherence of the RFI in the observed direction. However, RFI still has a deleterious effect, and strong RFI can ruin an observation even when it is received in only one of the array antennas. Other detection techniques for interferometric systems use the fact that interferometric radiometer observations produce a spatial covariance matrix whose elements consist of correlation products ("visibilities") between all pairs of antennas in the interferometer. Estimates of the number of RFI sources and their locations can, under certain conditions,

---

[35]Li Li, E.G. Njoku, E. Im, P.S. Chang, and K. St. Germain, "A Preliminary Survey of Radio-Frequency Interference over the U.S. in Aqua AMSR-E Data," *IEEE Transactions on Geoscience and Remote Sensing*, 42(2): 380-390 (February 2004).

[36]S.W. Ellingson and J.T. Johnson, "A Polarimetric Survey of Radio-Frequency Interference in C- and X-Bands in the Continental United States Using WindSat Radiometry," *IEEE Transactions on Geoscience and Remote Sensing*, 44(3): 540-548 (March 2006).

[37]C.A. Katz, J.N. Hewitt, B.E. Corey, and C.B. Moore, "A Survey for Transient Astronomical Radio Emission at 611 MHz," *Publications of the Astronomical Society of the Pacific*, 115: 675-687 (June 2003); and N.D.R. Bhat, J.M. Cordes, S. Chatterjee, and T.J.W. Lazio, "Radio Frequency Interference Identification and Mitigation Using Simultaneous Dual Frequency Observations," *Radio Science*, 40(5): 1-11 (June 2005).

be obtained from an eigenanalysis of this matrix. For example, strong RFI sources producing large eigenvalues can be detected in a manner analogous to the pulse detection process. Weaker RFI sources can be more difficult to detect, however. Interferometric detection strategies can be combined with the spatial excision suppression techniques discussed below.

Suppression Techniques

Suppression techniques can be divided into three categories: (1) *Filtering* is the simple process of designing receiver filters so that corruption from RFI sources outside a band of interest is minimized. (2) *Excision* is the removal of data in which RFI has been detected. A common property of all excision techniques is a partial loss of radiometry data as well as a possible distortion of non-excised radiometry data due to artifacts of the detection and excision process. (3) *Cancellation* is the subtraction of RFI from the radiometer output. Cancellation is potentially superior to excision in the sense that the RFI is ideally removed with no impact on radiometry, providing a "look through" capability that is nominally free of the artifacts associated with the simple "cutting out" of data. However, as discussed below, the trade-off with respect to excision is usually that suppression is limited. A further limitation of canceling techniques is that they tend to degrade into excision-type behavior when conditions are not optimal—for example, in low-interference-to-noise-ratio scenarios.

Filtering

While filtering is not necessarily a suppression strategy based on a detection process, its importance nevertheless motivates a brief discussion. All radiometry observations occur in a limited portion of the spectrum that is of interest for particular measurement purposes. Radiometer receivers are designed to include filter components to suppress the contributions of any emission sources outside the frequency range of interest. The performance of these filters can have a significant impact on the degree to which RFI corruption can occur. The bandpass filters used in radiometry ideally have a transfer function that is unity within the band of interest and zero outside this band, but such filters are not achievable in practice. Instead, the suppression of OOB power achieved by the filter "tails" must be traded against other filter performance properties. Strong RFI sources located outside the band of interest can make measurable contributions to radiometry measurements if filter performance is insufficient.

In terms of mitigation performance, it is desirable to place band-defining filter components as close to the antenna as possible, in order to reduce the tendency for strong out-of-band RFI to drive the receiver into nonlinear operation, resulting in compression or unacceptable intermodulation products. Unfortunately, analog filters are inherently lossy, so using such a filter degrades the sensitivity of a radiometer, presenting a difficult trade-off between basic performance and the ability to tolerate nearby out-of-band RFI.

Excision

Excision refers to the deletion of data that are believed to be contaminated by RFI. The use of excision implies that a data set is available from which some data are removed through a detection process and the remainder are used in estimating astronomical or geophysical information. Excision algorithms that have been explored to date utilize data sets based on measurements as a function of time, frequency, or space.

Temporal excision is the most common process. It is based on removing detected observations from a time series of measurements (in EESS applications, time-domain excision leads unavoidably to the excision of data corresponding to distinct locations as well). Temporal excision can be performed in conjunction with any of the detection algorithms described previously, and at time resolutions ranging from the Nyquist sample rate (i.e., nano- or microseconds) to post-integration timescales of seconds or larger. Temporal excision ensures that detected RFI makes no contribution to scientific analysis but at the same time reduces the amount of data available for the same analysis and potentially introduces artifacts that can compromise the value of the remaining data. The best case is that a reduction in the amount of available data merely reduces the sensitivity of the radiometer observation. Thus, it is desirable to implement temporal excision at a timescale that is comparable to that of any pulsed interference sources in order to retain the maximum amount of data. Temporal excision is best suited for sources that are localized in time and is often used with a pulse detection strategy. Numerous examples of temporal excision exist in the literature, and recent works have demonstrated real-time onboard temporal detection and excision. The performance of temporal excision in suppressing RFI source contributions is limited solely by the performance of the associated detection algorithm, which determines the false-alarm probability and probability of detection for specific RFI types.

When measurements in multiple frequency subchannels are available, RFI contributions detected in a particular subchannel can be removed by discarding data from that subchannel in computations of average powers or other averages across frequency. Frequency excision is limited to narrowband RFI, but it has the advantage (with respect to temporal excision) of being effective against persistent interference. In total power radiometry (most EESS observations and "continuum" RAS measurements), discarding data in a particular subchannel again has the effect of decreasing the sensitivity of the radiometer measurement when total channel quantities are of interest. This method is typically not acceptable in high-sensitivity spectroscopy, which is a commonly used mode in the RAS, although it is sometimes effective in imaging observations when the visibilities are computed on a narrowband basis. It is desirable to perform frequency excision at a frequency resolution that is comparable to that of observed RFI sources, so that a maximum portion of the noncorrupted spectrum can be retained. Given the fact that numerous RFI sources exist with bandwidths of 1 MHz or less, frequency resolutions below 1 MHz are desirable, but come at the cost of a greatly increased data rate if frequency excision is performed in postprocessing. Frequency excision has been demonstrated in combination with kurtosis, pulse, and narrowband detection strategies. Performance again is strongly dependent on the performance of the associated detection algorithm.

Spatial excision refers to the use of the beam-forming capabilities of compact antenna arrays—that is, arrays with maximum baselines on the order of wavelengths. One approach is based on synthesizing directly an antenna pattern null in the direction of a known RFI source; this is believed to be effective against RFI from satellites in RAS observations, although quite expensive and complex to implement. Sophisticated algorithms have been developed for this process in the RAS literature, and one of the key difficulties identified has been to minimize the impact of spatial excision on the main antenna lobe properties so as not to confound image calibration.[38] Spatial excision is further limited by the degree to which the array geometry and individual antenna patterns are able to generate deep nulls, and the number of such nulls that can be formed without unacceptable mainlobe and sidelobe distortion. While this technique is used frequently in military antijamming applications, the problem is more challenging for RAS observations due to the low signal-to-noise ratios of the astronomical signals of interest.

Suppression methods other than simple excision of RFI-contaminated data are not widely used in the EESS and RAS, mainly because they are not easy to devise or implement and may require the development of extensive special hardware, software, or instrument modifications that potentially degrade performance. Furthermore, cancellation techniques typically lead to significant increases in the required computing power relative to that needed in interference-free conditions. However, recent studies have developed and demonstrated cancellation approaches for RFI mitigation in RAS applications. Cancellation requires a detailed knowledge of the RFI signal—for example, a priori information about the modulation type, or a copy of the signal obtained by other means—in order to estimate and subtract its contributions to the data. Obtaining a precise description of source properties is difficult when the corrupting sources are observed at low instantaneous interference-to-noise ratios, as is the case for ground-based sources in the sidelobes of an upward-looking RAS antenna. Two strategies for improving RFI

[38] A.J. Boonstra, "Radio Frequency Interference Mitigation in Radio Astronomy," PhD Thesis, Delft University of Technology, Dept. EEMCS, June 2005, ISBN 90-805434-3-8.

TABLE 4.2 Successes and Limitations of the Unilateral Mitigation Methods Employed to Date by the Earth Exploration-Satellite Service

Type of Radio Frequency Interference (RFI)	Reference	RFI Details	Center Frequency (GHz)/ Bandwidth (MHz)	Detector Type	Detector Time/ Frequency Resolution
Pulsed	[1]	Out-of-band emissions from an ARSR system observed at close range	1.413/20	Kurtosis	36 ms/3 MHz
Pulsed	[2]	Out-of-band emissions from an ARSR system observed at close range	1.413/20	Pulse detection	10 ns/20 MHz
Pulsed	[3]	Unknown source of presumably out-of-band pulses	1.413/20	Analog pseudo-kurtosis	0.5 s/20 MHz
Pulsed	[5]	Airborne flight encountering many source types	1.413/20	Kurtosis	8 ms/20 MHz
Narrowband	[2]	Airborne test flight encountering many narrowband source types	5.5-7.7/100	Cross-frequency	26 ms/0.1 MHz
Wideband	[2] [4]	Airborne test flight encountering many wideband source types	5.5-7.7/100	Cross-frequency	26 ms/0.1 MHz
Wideband	[5]	Airborne test flights encountering many wideband source types	6, 6.4, 6.9, 7.3/400	Cross-frequency	26 ms/400 MHz
Gaussian-like		None	None	None	None

[1] C.S. Ruf, S.M. Gross, and S. Misra, "RFI Detection and Mitigation for Microwave Radiometry with an Agile Digital Detector," *IEEE Transactions on Geoscience and Remote Sensing,* 44(3): 694-706 (March 2006).
[2] B. Guner, J.T. Johnson, and N. Niamswaun, "Time and Frequency Blanking for Radio Frequency Interference Mitigation in Microwave Radiometry," *IEEE Transactions on Geoscience and Remote Sensing,* 45: 3672-3679 (2007).
[3] Jeffrey Piepmeier, Priscilla Mohammed, and Joseph Knuble, "A Double Detector for RFI Mitigation in Microwave Radiometers," *IEEE Transactions on Geoscience and Remote Sensing,* 46(2): 458-465 (2007).

Mitigation Type	Mitigation Time/ Frequency Resolution	Performance Achieved	Comments
Frequency subchannels	36 ms/ 3 MHz	Pulsed RFI ranging between 1-13 K in 20 MHz detected and removed (post-processing)	
Time blanking	40 microsec/ 20 MHz	Real-time removal of pulsed RFI ranging between 1-20 K in 20 MHz	
Time blanking	0.5 s/ 20 MHz	Pulsed RFI ranging between 1 K and 10-15 K in 20 MHz detected and removed (post-processing)	
Time blanking	8 ms/ 20 MHz	Pulsed RFI ranging from 0.1 to 45 K detected and removed (post-processing)	
Cross-frequency	26 ms/ 0.1 MHz	Narrowband RFI ranging from 1-45 K in 100 MHz detected and removed	
Cross-frequency	26 ms/ 0.1 MHz	Wideband RFI ranging from 10-100 K detected and removed	Removal of wideband RFI eliminates large portions of usable radiometer bandwidth; detection possible only when RFI power/MHz substantially exceeds that of noise.
Cross-frequency	26 ms/ 400 MHz	Wideband RFI ranging from ~5-300 K removed	Removal of wideband RFI eliminates large portions of usable radiometer bandwidth; detection possible only when RFI power/MHz substantially exceeds that of noise.
None	None	None	Not possible to detect RFI that resembles thermal noise.

[4] J.T. Johnson, A.J. Gasiewski, B. Guner, G.A. Hampson, S.W. Ellingson, R. Krishnamachari, N. Niamsuwan, E. McIntyre, M. Klein, and V.Y. Leuski, "Airborne Radio-Frequency Interference Studies at C-Band Using a Digital Receiver," *IEEE Transactions on Geoscience and Remote Sensing,* 44(7, Part 2): 1974-1985 (July 2006).

[5] A.J. Gasiewski, M. Klein, A.Yevgrafov, and V. Leuskiy, "Interference Mitigation in Passive Microwave Radiometry," *Proceedings of the 2002 International Geoscience and Remote Sensing Symposium,* Toronto, Ontario, Canada, June 24-28, 2002.

source knowledge are utilized. In the first, the upward-looking measurements of the RAS antenna are augmented with measurements from a "reference antenna" directed toward the source. This latter antenna observes RFI sources at a higher signal-to-noise ratio, which allows better estimation of RFI source properties in the cancellation process. A second approach is utilized for RFI sources with known modulations, for which a demodulation process can increase the signal-to-noise ratio. Given either a demodulation or second antenna measurement, cancellation then involves an estimation and subtraction of RFI source contributions to the data. The latter can be performed either "precorrelation" or "postcorrelation"—that is, either before or after the spatial covariance matrix is formed in an interferometric system. Cancellation performance is limited by the extent to which RFI sources can be detected and successfully estimated (a function of the signal-to-noise ratio at which the RFI sources are observed) as well as by the complexity and any temporal evolution of the RFI environment in which the observations occur.

Unilateral Mitigation Successes and Limitations

Table 4.2 above provides a short summary of the successes and limitations of the unilateral mitigation methods that have been employed to date by the EESS.

Finding: While unilateral radio frequency interference mitigation techniques are a potentially valuable means of facilitating spectrum sharing, they are not a substitute for primary allocated passive spectrum and the enforcement of regulations.

4.4 MITIGATION THROUGH COOPERATIVE SPECTRUM USAGE

The unilateral mitigation techniques described in §4.3 are at best a short-term solution to the RFI problem, which can be effective only when spectral occupancy is low and the RFI is easily distinguished from the background. This approach is otherwise inherently limited by the lack of coordination with the active services, and using only this approach, science users would perpetually be "guessing" how to work around RFI. This tactic will soon find its limits given the trends described in §4.1. A far more effective and efficient approach would be bilateral, or cooperative, mitigation.

Cooperative mitigation techniques would coordinate the timing and regional use of the radio spectrum in a far more dynamic manner than has existed with past technologies and regulatory structures. This is a new approach by which active services cooperate with passive (science) services within shared spectral bands by briefly interrupting or synchronizing radio transmissions to accommodate the science measurements. Such accommodations would occur only when and where those science observations are needed (e.g., during a satellite overpass), so the impact on spectrum availability for active services would be very low. The intelligence of modern devices makes this approach attractive, and there will likely be many instances where the costs of this mitigation technology would be readily accepted by users eager to gain access to large portions of the spectrum. Indeed, many devices will need to possess the necessary technologies and standards to negotiate spectrum use automatically among competing users, so the extension of these standards to accommodate science users could, in principle, be accomplished with very low costs and with a very low impact on functionality.

Cellular telephones provide a familiar example and some insight as to how this technology could work: cell phone networks automatically coordinate spectrum use among large quantities of transmitting devices. These systems provide a very dynamic command-and-control authority to assign frequency, or to interrupt or deny service, or to give priority (e.g., when a user dials 911) for each device within and among cellular regions. Conceivably, these systems already represent most—if not all—of the needed infrastructure for cooperative mitigation. The only missing elements are the agreed-on standards and the software that would allow these systems to momentarily relinquish assigned spectral bands in response to science requests. These could be communicated either directly from EESS satellites, for example, or from a networked database.

Consider the following scenario for cooperative mitigation. For this example, it is assumed that 30 spaceborne microwave radiometers are engaged in Earth observations for operational and research-oriented scientific purposes. This fleet of EESS satellites passes over a specific area several times per day but for only very brief intervals during each satellite's pass. The typical spot size of an EESS observation on Earth is about 30 km in diameter. Fixed or mobile transmitters operating within or near a receiving band used by the EESS could operate nearly full time if the transmitters were responsive to a blanking request signal or other preprogrammed transmitter time-off period that is coordinated with the overpass of each EESS sensor. Due to the brief time of footprint passage, this strategy would permit EESS receivers to measure microwave brightness temperatures while negligibly impacting active service performance. This would be especially helpful to EESS observations in bands that are not allocated to the service.

To determine the impact on active services, consider the fractional coverage of the fleet of EESS satellites. A "keep-out zone" of 10 times the footprint size, or 300 km in diameter, would generally ensure that the interfering source is well outside the near-sidelobes of the satellite instrument where it is most susceptible to interference. The total keep-out area on Earth for all spaceborne radiometers would then be of order $20\pi(150)^2 = 1.4$ million km^2, or an area of approximately 0.3 percent of Earth's total surface. If a random distribution of satellite locations is assumed, this fractional aerial coverage can, to first order, be equated with the fractional probability of occurrence of a satellite observation being made at a particular location on Earth. Put another way, an active user could, on average, transmit 99.7 percent of the time within the detection band of a passive satellite sensor without causing any EESS interference.

Another cooperative arrangement is illustrated by a hypothetical situation in which all Air Route Surveillance Radars (ARSR), which operate at L-band, would be synchronized to a time standard that allows them to be blanked for the approximately 20 milliseconds for several times per day that the radar transmitter is located within the moving antenna footprint of an overhead EESS sensor operating in L-band. The loss of information to the radar service would be minuscule, and given the ubiquity of modern Global Positioning System (GPS)-based time synchronization and Internet-accessible ephemeris data, the cost of the hardware and software necessary to perform blanking would also be small. However, the value of interference-free data to environmental monitoring and forecasting services at several critical EESS bands, specifically L-band, would be immense. Similar synchronization signals could be made available from registered transmitters (both fixed and mobile) to fixed RAS and EESS users or to other users of the spectrum who could then use them to blank their own observations or raise data quality flags. Blanking regions, in certain cases of strong transmitters, could need to be extended to take into account reflections from geographical features.

The above arguments and statistics strongly suggest that better time management of the available spectrum could result in significantly more time-bandwidth product being made available to passive services without impacting active services to any appreciable degree. To simplify the implementation, cooperative strategies are best implemented in bands used by fixed, registered transmitters—rather than unlicensed devices—although most new Internet-connected and GPS, cellular, or Wi-Fi devices could readily be required to contain simple software for cooperative mitigation. Cooperative mitigation techniques have been proposed over the past decade for commercial and consumer devices such as commonly used cordless telephones and devices for use in TV white spaces.[39] The extension of these techniques to the passive scientific community might provide many benefits. The committee anticipates that the active services could be viable partners in such an arrangement and could benefit from better usage of the active bands as well as from noteworthy public relations through their support of the EESS and RAS. It is conceivable, given appropriate management policy, that the passive spectrum could be "rented" to commercial interests when not needed, with revenues being used to support improved spectral usage studies and/or passive spectrum management needs.

Coordination between RAS ground stations and satellites containing trans-

[39]P. Kolodzy, M. Marcus, D. DePardo, J.B. Evans, J.A. Roberts, V.R. Petty, and A.M. Wyglinski, "Quantifying the Impact of Unlicensed Devices on Digital TV Receivers," Technical Report ITTC-FY2007-44910-01, University of Kansas, January 31, 2007.

mitters is critical for the present and future viability of the RAS, but it is much more difficult than ground-to-ground coordination, as is discussed in the next two paragraphs. For example, the coordination process between the RAS and Iridium, as discussed in §3.5, shows that coordination is not always successful at reducing RFI to needed levels.

As an example of successful collaboration, passive users of the spectrum and the Wireless Medical Telemetry Service (WMTS) were able to find a successful cooperative agreement in the 608-614 MHz in which both services still operate. In 1999, the U.S. Committee on Radio Frequencies (CORF) supported the FCC's proposal for RAS and WMTS to share this band as long as the proposal was enacted in its entirety to include "service rules on eligibility, frequency coordination with RAS facilities, the necessity to protect RAS observations from interference, and technical standards (including field strength, separation distance from the radio observatory, and out-of-band emission limitations)."[40] The proposal was enacted as supported by CORF, and the agreement between the services is seen by both parties as an excellent pairing of interests and one that has benefited them both substantially.

Similarly, a successful arrangement was made between the Arecibo Observatory and a nearby military radar station. The Puerto Rico Air National Guard operates a frequency-hopping radar with channels between 1220 and 1400 MHz at Punta Salinas, about 75 km from the telescope. Arecibo Observatory staff and the authorities at Punta Salinas devised a coexistence arrangement that involves blanking the transmitter when it is aimed at the observatory.

This is not meant to say that cooperative mitigation can replace the need for radio quiet areas or for restricted, passive-only bands. Indeed, since the development of passive techniques often occurs on unscheduled bases and in arbitrary regions, the need for emission restrictions within the small exclusively allocated spectrum and specific geographical zones remains. Many airborne and ground-based EESS experiments require continuous operation within a given zone and would not be able to yield effectively to active systems over time intervals exceeding even a few tens of percents. Such activity requires the use of restricted spectrum. Similarly, for the RAS, transmissions in geographical areas around radio telescopes must be avoided, and in order to maintain existing capabilities it should still be required that the RAS be given a chance to comment on all license applications for fixed and mobile transmitters within prescribed geographical zones around radio telescopes. However (and for example), in a shared time-of-day cooperative scheme, commercial traffic on certain shared bands of RAS frequencies might be

[40]National Academy of Sciences, "Comments on Docket No. ET 99-255, Amendment of Parts 2 and 95 of the Commission's Rules to Create a Wireless Medical Telemetry Service," filed with the Federal Communications Commission on September 30, 1999. Available at http://sites.nationalacademies.org/BPA/BPA_048830; accessed January 14, 2010.

acceptable in exchange for cooperative active access to other bands at suitable times, thus permitting effective radio astronomical observations to take place during transmission-free windows.

Finding: Nascent technologies exist for cooperative spectrum usage, but the standards and protocols do not.

The above finding is one of the key points of this section: the smart, inexpensive, portable, and highly networked electronics that are incorporated into many devices now have the capabilities needed for intelligent spectral sharing, but the organization of the manufacturing sector and the regulatory impetus needed to implement such sharing need to be developed jointly between the scientific and the industrial communities. It is likely that if such coordination can be developed, there will be additional spinoff benefits that will further facilitate spectral sharing within the purely active community as well.

4.5 MITIGATION COSTS, LIMITATIONS, AND BENEFITS

As the previous chapters and sections have illustrated, the nature of the costs of the interference problem for the EESS and RAS is wide ranging. The costs are manifested as impaired or even unusable data; costs are also incurred when the EESS and RAS programs must engineer technical or other fixes to mitigate the effects of interference on their operations.

Few of these costs can be monetized easily. The reason is that most of the value provided by the EESS and RAS is embodied in public goods—these include the host of environmental benefits and the improved ability to manage natural resources enabled by the EESS, and the enhanced or wholly new science understanding brought by the RAS. By definition, the societal benefit derived from public goods is difficult to express in dollar values. For example, even though improved forecasts are linked to reductions in weather-related loss of life and property, backing out the contribution of EEES data to this outcome is complex and difficult. It is even more complicated to back out from such a calculation the degradation associated with RFI.

This very problem is at the heart of spectrum allocation decisions when commercial services such as cellular telephones have an easily demonstrated market value, but scientific and other public uses of spectrum do not. As is well known from the literature on the value of public goods, however, simply because these goods are hard to monetize does not lessen their importance to society. Nor does this difficulty reduce the burden on decision makers to accord high importance to public uses in making resource allocation decisions such as those involving spectrum. In this chapter the committee has sought to inform these decisions by highlighting the costs of the interference problem for the EESS and RAS.

Earth Exploration-Satellite Service (EESS)

The challenge to the EESS below about 10 GHz is from interference arising from high-speed electronics that incidentally radiate isotropically (e.g., electronic cameras and computers), and from short-range wireless services such as Wi-Fi, Bluetooth, and cellular telephones. Interference above about 10 GHz arises from poorly filtered or directed communications, radar, and related services in bands in or near passive bands, or in bands with harmonics in passive bands (see Tables 2.1 and 2.2 and Figure 3.10). Equipment radiating above 10 GHz is mostly sold to large entities at prices well above consumer levels, and mitigating filters or other RFI suppression devices could readily be added to that equipment. One exception is automobile anticollision radar being developed for large-scale consumer sales for use in the 23 GHz band, despite that band's current worldwide exclusive ITU and FCC passive allocation (see also the discussion in §§2.5, 3.5, and 4.1). A potential future problem could arise if standards for widely used consumer equipment do not preclude incidental emissions above 10 GHz, which generally can be avoided with minor design changes at little cost.

Radio Astronomy Service (RAS)

The RAS is currently dominated by relatively few large radio observatories located in remote areas that nonetheless are beset by increasing levels of incidental interference from proliferating consumer-level electronics such as cellular telephones, Wi-Fi and Bluetooth systems, computers, and so on; from emissions from aircraft and satellites; and from over-the-horizon signals arising hundreds of miles away, well outside most protected areas but reflected by aircraft, the troposphere, and other means. Explicit expenditures for RAS mitigation research and implementation are modest because mitigation for the next generation of radio telescopes will be achieved primarily by the indirect costs of locating the observatories in extremely remote locations that are therefore more expensive to develop and operate (e.g., the Western Australian desert or the Chilean Andes). RAS costs are thus arguably already strongly affected by such remote-site mitigation costs, so little mitigation budget is left. Nonetheless, using horizon sensors to detect RFI of terrestrial origin is being pursued.

Nature of the Costs of Radio Frequency Interference to the EESS and the RAS

The discussion above illustrates that the costs of RFI to the EESS and RAS take several forms. One cost is the direct loss of information when RFI renders data and observations less useful or, in some cases, wholly unusable. This direct loss of information greatly reduces the societal value of the billions of dollars invested in the nation's EESS and RAS physical infrastructure.

Another cost is that of actions that must taken to accommodate RFI, provided accommodation is even possible. As discussed in Chapters 2 and 3, these actions include alterations in sensor deployment and operations, changes in scheduling, and other technical and engineering adjustments.

Examples of lost information content include many examples in the cases of both the EESS and RAS. In the case of the EESS these include the following:

- In some cases, despite extensive quality checking of EESS data, there are no good means of tracking the impact of a single observation that may be corrupted by noise. In the case of radiance assimilation for numerical weather prediction models, a single passive microwave satellite measurement that is contaminated with RFI at a level comparable to the satellite noise is unable to be distinguished from an uncorrupted measurement. The use of such a measurement can cause errors in an entire forecast.
- The direct measurement of water vapor and cloud water can be provided only by microwave radiometers. These measurements are commonly made in the 22-24 GHz frequency range, but microwave point-to-point communications and automobile anticollision radars operating in this spectral band are a source of significant RFI that will increase as automotive radar becomes more common.
- Another example is sea surface temperature, for which measurements are made at 10 GHz. Microwave brightness in littoral regions is impaired by contamination from the use of X-band spectrum adjacent to and within this spectrum allocation.
- The 10.7 GHz channels of AMSR-E are RFI contaminated over parts of Europe and Japan and are not used in these locations. (On a research basis, it is still possible to use the 6.9 GHz band for soil moisture retrieval over large regions.)
- RFI in bands below 10 GHz can compromise or even render unusable the unique soil moisture information obtained at 1.4 and 6 GHz.

In the case of the RAS, examples include the following:

- The detection of deuterium formed during the Big Bang and now found in interstellar gas was for many decades impeded by RFI. Detection was possible only after extensive shielding and the use of RFI monitors.
- To date, efforts to detect the redshifted (into the VHF-band) 1420 MHz emission of the epoch of reionization have been defeated by RFI; examples include experiments at Arecibo and at the VLA.[41,42]
- 1612 MHz imaging by the VLA was crippled by legal emissions from the Iridium satellite system until new filters were installed in the VLA. See Figures 3.10 and 3.11.

Characterizing these examples of lost information in financial terms is extremely difficult, given the public-good nature of the EESS and RAS. If a loss in the value of the information could be easily monetized, spectrum regulators would have some basis by which to compare the value of spectrum allocations to the EESS and RAS with allocations for consumer products that create many RFI problems. The methodological challenge posed by the comparison of public goods with consumer goods in deciding how best to allocate and manage spectrum among competing uses is well known.[43]

Another approach to characterizing the costs of RFI involves estimating the costs of the activities undertaken to mitigate or avoid RFI damage. In the case of the RAS, Box 4.1 describes some examples of mitigation costs. By asking what it costs to avoid or mitigate damage, regulators could compare the cost to the EESS and the RAS of avoiding damages with the cost to sources of RFI of mitigating their RFI (such as using filters or other means of RFI suppression). Whichever services face the least cost, either to avoid damage from RFI or to mitigate the creation of RFI, could be asked to bear the financial burden of taking the action. This approach of comparing costs can be useful to spectrum managers. However, because it only looks at costs, and not at benefits to society, of the information provided by the EESS and the RAS, the avoided-cost-based approach is inferior as a means of guiding spectrum management.

[41] J. Weintroub, P. Horowitz, I.M. Avruch, and B.F. Burke, "A Transit Search for Highly Redshifted HI," Astronomy Society of the Pacific Conference Series, Vol. 156 (1999).

[42] Greenhill, L., et al., "Mapping HI Structures Present During the Epoch of Reionization," IR&D Report, Center for Astrophysics, Harvard University.

[43] Harvey J. Levin, *The Invisible Resource: Use and Regulation of the Radio Spectrum*, Baltimore: Resources for the Future and Johns Hopkins University Press, 1971.

BOX 4.1
Illustrations of Radio Astronomy Actions and Costs for Radio Frequency Interference Mitigation

- *Example 1: Using knowledge of the local environment.* Experts use patterns such as the time of day or year to identify local sources of radio frequency interference (RFI), which can range from a lawn mower to an Iridium-based aerostat used by police for surveillance. In these cases, RFI is solved by coordination. Tracking down the RFI source typically uses about 1 full-time-equivalent (FTE) day to solve.
- *Example 2: Sleuthing with radio direction finding equipment.* This procedure requires an RFI van equipped with receivers, amplifiers, a spectrum analyzer, and a directional antenna—equipment that costs approximately $20,000. In these cases, which are infrequent, RFI is solved by coordination. Here, tracking down RFI may require 2 to 3 FTE days.
- *Example 3: Tracking ambiguous external RFI.* Tracing RFI to a specific satellite source can be time-consuming and difficult. It may also have legal ramifications. These problems can take an FTE month and require archive work, software development, and detailed knowledge of the satellite (system specifications, operating parameters) that may be the source of interference.

Summary

Increasing levels of incidental interference from proliferating consumer electronics and other sources threaten the EESS and RAS. The primary current EESS problem is active services in passively allocated bands where the atmosphere is sufficiently transparent that EESS instruments see Earth's surface. The RAS is strongly affected by emissions from aircraft, satellites, and over-the-horizon signals, necessitating the siting of sensitive observatories in remote locations. All RFI poses the potential for loss of information in EESS and RAS observations and data, thus undermining realization of the full societal benefit of Earth and radio astronomy science.

5

Findings and Recommendations

The allocation and use of radio frequencies constitute a complex issue at the center of many different fields of inquiry, from engineering to economics. The committee was tasked to "prepare a report exploring the scientific uses of the radio spectrum which will:

- Portray the science that is currently being conducted using the radio spectrum;
- Identify the spectrum requirements necessary to conduct research;
- Identify the anticipated future spectrum requirements for at least the next 10 years; and
- Advise spectrum policy-makers on the value to the nation of accommodating scientific uses of the spectrum, recognizing the need to balance multiple communities."

The committee chose to focus its efforts on the passive uses of the spectrum, primarily in Earth remote sensing and radio astronomy. The committee recognizes that there are many other scientific uses of the spectrum, but it focused on the passive uses because these activities pose unique challenges to the nation's spectrum allocation and management policies.

During the course of the study, the committee identified a number of key findings and formulated recommendations concerning passive uses of the radio spectrum for scientific purposes over the next two decades. The findings identify the operational and educational value of these uses to the broader society as well as describe the rapidly developing threats to the viability of some areas of this science as a result of increasing use of the spectrum by active systems. Active use of the spectrum has in and of itself generated unprecedented degrees of economic prosperity, enlightenment, and security. Although the pressure for active use of the spectrum cannot and should not be reduced, the committee nonetheless identified a number of measures that can be taken to help ensure the viability of the passive uses. The recommendations stemming from the committee's study and the findings on which they are based are presented in this chapter.[1]

5.1 SOCIETAL VALUE OF THE PASSIVE SERVICES

In addressing the first and fourth bulleted items in its statement of task, the committee focused on the purpose of the various passive applications within the Earth Exploration-Satellite Service (EESS) and the Radio Astronomy Service (RAS), and on how these purposes align with societal needs. A wide range of passive applications exist in Earth remote sensing and radio astronomy that facilitate day-to-day environmental services, scientific inquiry into basic physics and environmental processes, and both formal and outreach education. The committee expects that the societal value of the passive services will grow in importance over the next two decades.

Passive microwave remote sensing observations provide a valuable and important set of tools that contribute to the monitoring, understanding, and predicting of the many key components of Earth's natural system and that are essential for the understanding of the interaction of these components so that weather and climate can be analyzed and predicted. Passive EESS measurements are increasingly used directly in numerical environmental models that help predict weather and analyze global climate change. These observations represent the only viable means by which certain key environmental parameters can be measured. They are a cornerstone of the ability of the United States to maintain its preeminence in Earth science and are critical to the economic vitality and health and safety of the nation's people.

Finding: Passive remote sensing observations are essential for monitoring Earth's natural systems and are therefore critical to human safety, the day-to-day operations of the government and the private sector, and the policy-making processes governing many sectors of the U.S. economy.

While Earth scientists study the natural radiation from Earth and the atmosphere, radio astronomers use similar techniques to study the natural radiation

[1]The committee's findings, as originally presented in the context of the discussion in Chapters 1 through 4, are reiterated here.

of sources in space. Radio astronomy has made fundamental contributions to the understanding of the nature, origin, and evolution of the universe, galaxies, stars, and planets.

Finding: Radio astronomy has great potential for further fundamental discoveries, including the origins and evolution of the universe, the nature of matter, and life in other solar systems, which will have an enormous impact on our understanding of fundamental physics and the place of humanity in the universe.

Radio astronomers and remote sensing scientists often push the state of the art in system design, leading to new developments in advanced signal processing, low-noise receivers, and novel antenna designs, among other areas. Computer algorithms developed for the RAS and EESS have also found routine application in medical imaging.

Finding: In addition to the intellectual benefits that they provide, radio astronomy and passive microwave remote sensing studies provide many technological benefits to American society.

Radio astronomers have produced many opportunities for scientific and engineering education, ranging from the K-12 through graduate levels. Scientific results from radio astronomy and Earth remote sensing continue to capture the imagination of the public, which is excited and awed by new discoveries about the universe and concerned about extreme weather events and possible climate change. Public interest is reflected in the large numbers of people who visit radio observatories every year and regularly follow weather and climate forecasts. The development of passive microwave sensors for both the EESS and RAS also provides an important training ground for the next generation of radio scientists and engineers.

Finding: Radio astronomy and passive microwave Earth remote sensing provide a diverse and valuable set of educational opportunities.

The federal government has historically recognized the importance of both of these fields to the nation. One measure of that recognition is the level of resources that the nation has invested in these endeavors. Fulfilling the scientific promise of radio astronomy and Earth remote sensing has required investment in a diverse group of observatories, sensors, and instrumental capabilities. Further progress in environmental modeling and forecasting, astronomy, and related areas of physics depends on continual improvements in the sensitivity of radio telescopes and passive microwave sensors on surface-based, airborne, and spaceborne platforms.

Finding: Scientific advances have required increasing measurement precision by passive radio and microwave facilities in order to obtain more accurate and thus more useful data sets. This need for precision will continue to increase.

Finding: Large investments have been made in satellite sensors and sensor networks and in major radio observatories. New facilities costing billions of dollars are under construction or are being designed.

Recommendation 1: Recognizing that the national investment in passive radio astronomy and Earth remote sensing is dependent on access to the radio spectrum, the committee recommends that the Federal Communications Commission (FCC) and the National Telecommunications and Information Administration (NTIA) ensure that access to spectrum for passive radio and microwave observations of Earth environmental variables and radio astronomical observations of the sky is protected in the development of future spectrum policy.

5.2 CHARACTERISTICS OF THE PASSIVE SPECTRUM SERVICES

The committee noted the following broad characteristics of passive EESS and RAS activities and applications. RAS and passive EESS equipment receive natural radio emissions from space or Earth (respectively) and use no transmissions. Accordingly, they do not cause radio frequency interference (RFI) to any other service. The signals received from cosmic or natural terrestrial sources are typically far weaker than the internal noise levels of the receivers. The required sensitivity of RAS and EESS systems is determined by the natural, minute level of radio emissions. Spectral band needs are determined by basic physical processes, and many measurements require spectrum at specific frequencies set by the spectral lines from the quantum transitions of atoms and molecules. These characteristics of RAS and EESS systems are likely to remain true over the next two decades, as they are intrinsic to the conduct of these activities. However, unmet spectral allocation requirements exist.

Finding: Effective passive microwave band allocations are necessary for the performance of environmental and radio astronomy observations.

Finding: Owing to their receive-only nature, the passive Earth Exploration-Satellite Service and Radio Astronomy Service, operating from 10 MHz to 3 THz, are incapable of interfering with other services.

Finding: Radio wave bands (10 MHz to 3 THz) are indispensable for collecting information associated with specific astronomical and environmental phenomena.

Often the same bands are similarly indispensable for both passive Earth remote sensing and radio astronomy, and the passive nature of both services enables them to share the spectrum productively.

The preceding findings have a number of important implications with respect to how radio astronomy and Earth remote sensing are currently conducted. Since the science requirements drive the need for additional bands and bandwidth beyond those allocated to the services, the RAS and EESS communities routinely use spectrum beyond these allocations on a non-interference basis, and with varying degrees of success. Such opportunistic sharing is essential for certain scientific measurements; it requires the careful design of experiments to avoid RFI.

Whereas technological advances have rapidly increased the channel capacity of spectrum available to active users, the same cannot be said for the passive services: they cannot use their allocated spectrum more efficiently. For instance, the passive services cannot use coding and compression techniques to expand the capacity of this bandwidth. Passive microwave sensors rely on their entire allocated bandwidths, and often much more, to achieve required measurement precisions.

Finding: Currently, 2.07 percent of the spectrum below 3 GHz is allocated to the RAS and EESS on a primary basis, and 4.08 percent is allocated on a secondary basis (measured in hertz).

Debilitating postlaunch RFI occurred in one major international passive environmental sensor mission at C-band (Advanced Microwave Scanning Radiometer-Earth [AMSR-E]), rendering soil moisture measurement impossible over several populated areas. Such RFI also occurred at C-band in a non-mission-critical manner in another U.S. passive microwave military sensor (WindSat). A spectral allocation at C-band is currently required for observations of soil moisture and sea surface temperature, and a wider allocation at X-band would be valuable for observing ocean wind direction. While, the spectral band from 10.6 to 10.8 GHz is still relatively free of RFI over the United States, growth in use of this band and C-band by active applications is anticipated.

Finding: There is currently inadequate protected spectrum in C-band and X-band for operational passive microwave observations of sea surface temperature, soil moisture, and ocean surface wind speed and direction.

A further characteristic of EESS measurements is that they are made on a continuous and global basis. Passive microwave and millimeter-wave sensor beams pass approximately 20 times per day over a typical location in the United States.

Because there is no EESS allocation within C-band and this portion of the spectrum is heavily utilized, measurements of brightness temperature at C-band over land are currently only considered observations of opportunity. RFI in this band not only biases measurements but causes observation failure. Global protection is needed due to the band's wide application in observing sea surface temperature, soil moisture, and ocean surface wind direction—elements critical to the understanding and predicting of Earth's environment.

Recommendation 2: The FCC and NTIA should move toward developing a passive EESS reference band allocation within 6-8 GHz to facilitate unilateral RFI mitigation. To be effective, this band should be at least 20 MHz wide and should be established on a global basis.

Such a reference band allocation would benefit radio astronomy as well. It would be advantageous for the RAS if this band included the methanol transition line, for example, which provides strong maser emission from star-forming regions in the Milky Way.

Finding: Whereas most frequency regulations for active services are defined on local or regional bases, passive EESS observations are global by nature. As a result, a high level of international cooperation is required to maintain and enforce passive allocations.

Recommendation 3: The United States should actively engage the international community on passive EESS and RAS frequency allocations in order to improve the availability of global measurements of environmental variables and radio astronomy observations.

5.3 THREATS TO THE EESS AND THE RAS FROM UNINTENTIONAL RADIO FREQUENCY INTERFERENCE

The radio environment in the United States and around the globe is rapidly changing due to the proliferation of active devices. This trend is likely to continue in the foreseeable future. It threatens the ability to use the spectrum for passive scientific purposes through inadvertent radio frequency interference. The committee assessed both the current state of the occurrence of RFI to the passive services (Chapters 2 and 3) and trends in spectrum usage (Chapter 4).

The most salient change in the use of the radio spectrum over the past 20 years has been the explosive growth in commercial use of the spectrum. Active commercial use of the spectrum will continue to grow in the number of links (2 billion or more cellular telephone users plus many additional data networks), the modes of usage (including data, voice, and active sensing applications), and geographic deployment (including near-rural and rural environs). These devices will be highly mobile, will use more of the spectrum, and will extend to geographic locations previously considered to be radio-quiet.

Finding: Radio frequency interference threatens the scientific understanding of key variables in Earth's natural system, now and in the future.

Weak cosmic signals of fundamental importance to radio astronomy are easily masked by human-made radio emissions. Even signals far below the sensitivity of high-quality receivers used by the active services can interfere with routine astronomical observations.

Finding: The emergence of practices for the dynamic use of the spectrum will result in more active devices with greater variability in active spectrum usage, and the EESS and RAS communities could be impacted with more unintentional radio interfering devices.

The proliferation of wireless devices and high-speed digital radio technology around the globe diminishes the value of Earth observations from remote sensing platforms, leading to an irrevocable loss of environmental data. When affected by RFI, EESS observations have increased potential for introducing errors in environmental forecasts on both local and regional bases.

Finding: Geographical separation of radio telescopes from transmitters (e.g., through the establishment of radio quiet zones and the remote siting of observatories) is currently effective in avoiding much radio frequency interference, but the proliferation of airborne and satellite transmissions and the widespread deployment of mobile, low-power personal devices threaten even the most remote sites.

Unlike Earth remote sensing applications, which require global coverage, radio astronomy has historically taken advantage of the benefits provided by geographical separation; large observatories have been built in remote, largely radio-quiet areas.

Finding: Important scientific inquiry and applications enabled by the Earth Exploration-Satellite Service (EESS) and the Radio Astronomy Service (RAS) are significantly impeded or precluded by radio frequency interference (RFI). Such RFI has reduced the societal and scientific return of EESS and RAS observatories and necessitates costly interference mitigation, which is often insufficient to prevent RFI damage.

Despite these findings, the current knowledge of actual spectrum usage is inadequate to address RFI threats to the EESS and RAS. The federal government collects more information about many other economic variables than it does for the current usage of the radio spectrum. A monitoring capability would aid in both mitigation and instrument design and in the identification of dynamic sharing opportunities. This information would also aid in enhancing current passive radio science as well as aiding the expansion of current EESS and RAS capabilities.

Finding: Greater efforts to collect and analyze radio emission data are needed to support the enforcement of existing allocations and to support the discussion and planning of spectrum use.

Finding: Better utilization of the spectrum and reduced RFI for scientific as well as commercial applications are possible with better knowledge of actual spectrum usage. Progress toward these goals would be made by gathering more information through improved and continuous spectral monitoring. This would be beneficial to both the commercial and the scientific communities.

Recommendation 4: The Department of Commerce/National Telecommunications and Information Administration (NTIA), in collaboration with the National Science Foundation (NSF), NASA, and the National Oceanic and Atmospheric Administration (NOAA), should spearhead the development of a national spectrum assessment system that measures the radio frequency (RF) environment with appropriately high resolution in time, space, and frequency for purposes of spectrum development and management, based on the spectral and spatial density of emitters.

The assessment of spectrum usage needs to occur at time, space, and frequency scales commensurate with actual usage. To this end, 1 microsecond (μs) would resolve many pulsed radar applications, and 1 kHz would be sufficient to separate and identify almost all individual signals in bands above and below 30 MHz for both voice and data. Spatial and angular resolution requirements are more difficult to identify. The necessary angular resolution would be frequency-dependent, such that the survey would achieve lower resolution at lower frequencies and higher resolution at higher frequencies. Since different communications systems use a very wide variety of spatial scales, finding a single spatial resolution necessary to conduct a useful survey is impossible; it comes down to what can be afforded. Crucially, however, all of these measurements should be of sufficient resolution to determine the adverse impact of most radio transmissions on the RAS and EESS. Spectrum monitoring with these guidelines would provide both statistical and operational information for the RAS and EESS, as well as providing many ancillary benefits to other scientific, commercial, and government applications and services.

5.4 TECHNOLOGY FOR MITIGATION OF RADIO FREQUENCY INTERFERENCE

Given the increasing threat to the passive uses of the spectrum posed by human-made transmissions, the RAS and EESS communities have studied the potential for the mitigation of unintentional RFI on both unilateral and cooperative bases. Bilateral mitigation technologies could potentially lead to effective spectral sharing between the active and passive services and could be particularly valuable for facilitating passive observations in non-allocated bands. The following findings and recommendations pertain to the current and projected future status of unilateral and cooperative RFI mitigation strategies suitable for maintaining the ability to use the spectrum for passive scientific purposes.

Finding: While unilateral radio frequency interference mitigation techniques are a potentially valuable means of facilitating spectrum sharing, they are not a substitute for primary allocated passive spectrum and the enforcement of regulations.

Techniques for the excision or subtraction of RFI continue to be developed, but they are only partially successful. For example, unilateral RFI mitigation techniques for passive EESS systems have been and continue to be explored. Only limited reports of success are available, however, especially with regard to levels of RFI comparable to the system sensitivity. Radio astronomy currently makes use of bands allocated to other services but sometimes is faced with the need for RFI mitigation. No set of universally effective techniques has been identified. Unilateral RFI mitigation could be facilitated by improved a priori information (e.g., time-space-frequency-angle structure) on spectrum usage, as recommended in Recommendation 4 in the preceding section.

Recommendation 5: The National Telecommunications and Information Administration (NTIA) and the appropriate National Science Foundation (NSF) and NASA units should promote the development of inexpensive out-of-band interference mitigation technology and testing capabilities (e.g., filters, modulation techniques, etc.) that could be added to and required for type-approved consumer devices for the protection of EESS and RAS bands. As these technologies become affordable, the technical regulatory rules should reflect these new capabilities.

Supporting the development of mitigation technology for application to the appropriate future radiating devices could preempt much interference to the passive services. As the technology matures and cost falls, the efficacy and availability of the technology should be reflected in regulations moderating spectrum use.

Recommendation 6: Investment in the development of mitigation technology should be increased so that it is commensurate with the costs of data denial that result from the use of systems without mitigation. To this end, NSF and NASA should support research and development for unilateral[2] RFI mitigation technology in both EESS and RAS systems. NASA, NOAA, and the Department of Defense should require that appropriate RFI analyses and tests and practical RFI mitigation techniques be applied to all future satellite systems carrying passive microwave sensors.

A secondary benefit of such research would be to quantify the qualitative and limited documentation of unilateral RFI mitigation capabilities and their ultimate utility, as well as to help quantify spectrum usage. The committee believes that an effort of several million dollars per year over 5 years could yield substantial results in this area.

Finding: Nascent technologies exist for cooperative spectrum usage, but the standards and protocols do not.

Cooperative spectrum usage is potentially more useful than is unilateral RFI mitigation, but the requisite ability to assign spectrum usage dynamically is currently undeveloped. Anticipating that the commercial, military, and scientific uses of the spectrum will continue to grow, there will be a commensurately growing need to cooperate on the usage of spectrum. Spectrum is underutilized over time, space, frequency, and angle, and cooperative spectrum usage offers a means of taking advantage of this underutilization.

One example of cooperative spectrum usage is time-domain multiplexing of spectrum over broad bandwidths. In such a scheme, the EESS or RAS would have exclusive use of spectrum for certain intervals, while tolerating transmissions from active services during the remaining time. This technique would be one way in which the anticipated evolution of spectrum utilization and management could result in a mutually successful scenario for both passive and active services, albeit with some increase in the complexity of equipment. The anticipated technical requirement is similar to proven existing technology that facilitates the time-division-multiplexed use of spectrum in cellular telephone systems such as the Global System for Mobile communications (GSM). However, since RAS experiments usually require a fixed integration time to achieve statistical accuracy, a time-domain multiplexing sys-

[2]As discussed in Chapter 4.

tem would increase the actual time per experiment and further strain the heavy demands on all of the world's large radio telescopes.

Recommendation 7: The NSF, NASA, and NTIA should jointly support research and development for cooperative RFI mitigation techniques and the associated forums and outreach necessary to enable the development of standards for greater spectral utilization and interference avoidance.

The committee believes that an effort of several million dollars per year over 3 years would be sufficient to demonstrate core technologies and to develop an implementation roadmap for these technologies. One end goal of these efforts would be to enable dynamic spectrum-sharing technology that would facilitate the observation of astrophysical phenomena which require measurements over large fractional bandwidths, such as observation of the redshifted 21 cm emission from the universe's Dark Ages and the epoch of reionization, pulsars, single pulses hypothesized to be associated with prompt emission from gamma-ray bursts and other extreme astrophysical phenomena. Such measurements are extremely difficult to make at this time and would provide effective benchmarks for the success of cooperative spectrum-sharing techniques. A moderate portion of this investment should justifiably be spent on informing the public and the relevant scientific and technical communities about EESS and RAS requirements, mitigation needs, and capabilities, and on the development of standards.

Recommendation 8: As cooperative spectrum-sharing techniques come into use, NSF and NASA spectrum managers should work with the regulatory agencies to enable observations that require an extremely wide spectral range. Such observations would provide a useful metric for the effectiveness of spectrum-sharing techniques for the passive services.

5.5 PROTECTION OF THE EESS AND THE RAS

The committee discussed at length actions that should be undertaken by U.S. agencies to ensure the continued benefits to the public of the passive services. Some of these actions can be undertaken solely within the United States and others would require international collaboration. The committee considered the costs and complexity versus expected benefits to the passive services carefully. In some cases, the committee identified existing ambiguities in rulemaking that could lead to an eventual loss of utility of the primary passive bands, thus warranting a clarification of existing regulations. In other cases, more complex regulatory measures must be undertaken to ensure the viability of the passive services.

One example of such an ambiguity involves the differences between the International Telecommunication Union (ITU) and the Federal Communications Commission (FCC) regulations in out-of-band and spurious emissions. In some cases, emissions that create harmful levels of interference are currently permitted in EESS and RAS primary bands, although the ITU regulations state that "all emissions are prohibited." The FCC regulations do not allow any primary emission, but out-of-band and/or spurious emissions from other bands are permissible. Thus, a device can meet the specific emission requirements and emit into the protected EESS and RAS primary bands. In order to protect primary EESS and RAS bands adequately, it should be required that out-of-band and spurious emissions be significantly attenuated when they fall within EESS or RAS primary bands. This may require the reconsideration of many of the emission limits of bands that are spectrally close to the EESS and RAS primary bands, for modification of the permitted OOB and spurious emission levels.

Finding: The rules for out-of-band and spurious emissions in the primary allocated Earth Exploration-Satellite Service (EESS) and Radio Astronomy Service (RAS) bands (e.g., 1400-1427 MHz) do not provide adequate interference protection for EESS and RAS purposes.

The rules that pertain to the above finding are given in Appendix D.

Recommendation 9: The NTIA and FCC, with the support of the NASA and NSF spectrum managers, should study rulemaking changes that require aggregate emission protection and out-of-band and spurious noise protection in primary EESS and RAS bands.

More complex methods of understanding and managing spectrum usage may also be required to enable more efficient spectrum usage.

Finding: Current regulatory structure and support infrastructure (such as databases, etc.) are transmitter-centric. Methodologies to incorporate passive systems need to be developed.

Finding: New cooperative spectrum management techniques that could be beneficial for enhanced interference management and increased spectral utilization have been investigated by regulators but have not been implemented.

The current regulatory structure inhibits the distribution of critical information on how active systems can impact passive systems, and it also limits promotion of the communications needed between active and passive users to enhance channel capabilities and limit inadvertent RFI.

Beneficial cooperative spectrum management techniques include the use of interference metrics (e.g., interference temperature), the extension of enforcement technology (e.g., the development of commercial devices used for enforcement measurements and additional mobile measurement systems), and the inclusion of passive systems in regulators' databases (e.g., the FCC's Universal Licensing System).

Recommendation 10: FCC and NTIA regulators should actively define interference metrics, expand enforcement technology, and include descriptions of passive EESS and RAS systems in regulators' databases.

However, many of the current gaps in the regulatory system stem from a lack of communication between—or even within—the various user communities. For example, there is currently no forum in the United States for identifying EESS frequency allocation needs and vetting the merits of alternative allocations within the context of all competing services (both public and private). To engender the requisite communication, the committee makes the following recommendations.

Recommendation 11: The EESS and RAS communities should be provided additional support through NSF, NASA, and NOAA to increase their participation in spectrum management forums within the International Telecommunication Union (ITU), FCC, NTIA, and other organizations. The goal of such participation is to foster outreach, advance the understanding of interference and regulation issues, and initiate mutual cooperation in interference mitigation.

For example, NASA and the National Oceanic and Atmospheric Administration (NOAA) could jointly sponsor a workshop to explore alternative means of addressing RFI, seeking participation from the FCC, NTIA, industry, vendors, and the university community. This workshop could focus on the development of satellite and aircraft payloads and ground-based systems that characterize spectrum use. Such an endeavor would help determine the need for modified and/or tightened regulations and would increase the general level of understanding about interference. The planning of this workshop could be facilitated by professional societies already engaged in similar activities.

Recommendation 12: The Office of Science and Technology Policy should create a new, permanent, representative technology advisory body to identify technical and regulatory opportunities for improving spectrum sharing among all active and passive users, both government and nongovernment.

The advisory body recommended here should include representatives from all user and regulatory sectors. In a common forum, these representatives could bring to bear the technical and regulatory creativity, breadth, and depth necessary to identify and ensure that new opportunities for improving spectrum sharing and utilization are brought in a timely way to the attention of the many existing, relevant, government and private bodies that now separately address more limited and immediate spectrum issues.

In addition to expanding communication, it is important to adjust the regulatory process in such a manner as to discourage new instances of unintentional RFI from arising in the future.

Recommendation 13: The FCC and NTIA should require active service users to use their allocated portions of the spectrum more effectively. Spectral efficiency requirements should be built into FCC and NTIA licensing policies for future spectral assignments.

Although out-of-band emissions restrictions apply to individual devices and these restrictions generally preclude RFI by an individual device, there is currently no way to ensure that when such devices are sold, the aggregate emissions from a large number of them will not cause harmful RFI. Limitations on aggregate emissions may be difficult to enforce, but the likelihood of RFI can be minimized prior to the sale of devices by considering assessments of realistic market penetration and usage concentration when emissions standards are being developed.

Recommendation 14: NASA, NOAA, NSF, and other agencies with interests in the EESS and RAS should oppose all type-approval licenses for equipment without source mitigation that impacts EESS and RAS bands.

Recommendation 15: A combination of radio impact statements and/or statements of compliance with interference mitigation standards and emission standards should be mandated to accompany all proposals to federal agencies for the research and development of active service technology.

Recommendation 16: The FCC and NTIA should follow up on specific recommendations of the U.S. Spectrum Policy Task Force (November 2002) to encourage spectral efficiency, maintain EESS and RAS spectral allocations, and be prepared to enforce spectrum protection.

In its final report, the FCC's Spectrum Policy Task Force made specific recommendations including the following: (1) ensure that the FCC has sufficient resources to independently monitor and enforce spectrum management rules,

(2) improve the out-of-band interference performance of transmitters and receivers, (3) adopt a standard method for measuring the noise floor, (4) create a public/private partnership for a long-term noise-monitoring network and for the archiving of data for use by the FCC and the public, (5) promote transmitter enhancements for interference control, (6) study the tightening of out-of-band emission limits, (7) accompany a clearer definition of interference with effective enforcement, (8) develop technical bulletins that explain interference rules for all radio services, and (9) develop the opportunistic or dynamic use of existing bands through either cognitive radio techniques to find white space in existing bands or use protocols to relinquish bands to primary users.[3]

5.6 THE PATH FORWARD

The radio spectrum is a finite resource that has been managed as such for the past 70 years by the federal government. This management enabled the growth of strong commercial and scientific communities. The pursuit of better techniques to leverage the unique characteristics of the radio spectrum has led to discoveries and innovations of enormous scientific and societal value. Over the past 20 years, rapid technological improvements have increased the capabilities of both the scientific uses and commercial uses of the radio spectrum exponentially. The current regulatory regime is struggling to enable the capabilities and desires of either of these communities, let alone both. A new path is needed to maintain the vital engines both for the scientific discoveries that lead to societal benefit and for the commerce that is straining to meet the demands of a mobile society.

Technological innovations continue to increase the utility of the radio spectrum. The onset of new technologies designed to exploit the diversity of the radio spectrum in space, frequency, polarization, and time will increase the efficiency of its use. However, the current means of managing spectrum use must be changed, as the current policies threaten to thwart scientific discovery, diminish the utility of important environmental observations, and limit economic growth. Therefore, new spectrum management policies need to be explored to foster these critical national capabilities.

The next generation of spectrum management policies must enable better sharing of the spectrum as well as contribute to a full understanding of the actual use of the RF spectrum. This can be done by exploiting currently available technologies and hastening the development of nascent technologies. New policies should encourage the following:

[3] Federal Communications Commission, *Report of the Spectrum Policy Task Force*, Washington, D.C., November 2002.

- *The development of the means for direct interaction between the active and passive spectrum users in order to protect current and future scientific uses of the spectrum.* The nation needs to provide the policies to strike a balance between pursuing advanced technology to decrease the cost of communications on the one hand and, on the other, to make the spectrum more usable and less noisy for all users;
- *A regulatory environment that enables sharing the spectrum in both space and time.* This is a "win-win" scenario that will enable additional scientific uses without impacting commercial development; and
- *Investment in technology to enable spectrum sharing between active and passive users, over the entire radio spectrum.* This investment should become commensurate with the investment made in remote sensing technology.

In one sense, the management of the spectrum for passive purposes can be likened to the management of U.S. public parklands. While monetization of the spectrum by the free market may be one value metric, the true societal value of EESS and RAS spectrum should more properly be assessed in a manner consistent with how public parklands have been valued. As history continues to show, parklands reserved for public enjoyment, with limited to no development being permitted, have a high intrinsic community value. Humankind has ultimately found a compelling need for such land, and this need has resulted in the preservation of parcels even within the most crowded urban areas where these parcels would otherwise sell on the open market at a premium price. There is a balance between development and preservation that recognizes the intrinsic value of parklands.

More often than not, the very presence of such public land increases the value of adjacent private land beyond the value that it would otherwise have. In a similar manner, the EESS and RAS studies performed using passive spectrum often lead to improved communications technologies and scientific insights that engender efficiencies and hence enhance profits and improve services within the private and public sectors.

The new initiatives necessary for spectrum management and sharing will not be easy to design and implement, nor will they make successful management and sharing a certainty. It will likely take a national effort to understand clearly the needs of both communities, the scientific and the commercial, and to motivate each to make the choices necessary to enable greater access for each to the radio spectrum. That said, it should be clear that the next generation of scientific users of the radio spectrum needs to be afforded the capacity to develop the technology that will open new horizons.

5.7 CONCLUSION

The passive services both provide a critical return to society through operations in support of environmental prediction and offer scientific intellectual value. The impact of the latter is difficult to quantify, but it has been seen to make unique contributions to our nation's progress. It would thus be in the strongest interests of the nation to see that access to spectrum for scientific purposes is maintained during the coming decades. The committee's recommendations provide a pathway for putting in place the regulatory mechanisms and associated supporting research activities necessary to accomplish this important task. The committee believes that such a pathway will also lead to greater efficiency in the active use of the spectrum, which should benefit all direct and indirect consumers of wireless telecommunications and data services.

Appendixes

Appendix A

Statement of Task

The current system of allocating bands in the radio spectrum was developed more than 50 years ago, and a review of the needs of scientific users is in order. In recent years, the explosion of new wireless technologies has significantly increased the demand for access to the radio spectrum. The increased demand has led to discussions in both government and industry with respect to new ways of thinking about spectrum allocation and use. Scientific users of the radio spectrum (such as radio astronomers and earth scientists using remotely sensed data) have an important stake in the policies that will result from this activity. It is proposed that a survey of the scientific uses of the spectrum be conducted to identify the needs of today's scientific activities and to assist spectrum managers in balancing the requirements of the scientific users of the spectrum with those of other interests. The survey will be carried out by a National Research Council (NRC) committee over a period of 18 months.

A balanced committee of 15 people will be formed to prepare an NRC report surveying scientific uses of the spectrum. Following is the committee's statement of task:

> The committee will prepare a report exploring the scientific uses of the radio spectrum which will:
>
> - Portray the science that is currently being conducted using the radio spectrum;
> - Identify the spectrum requirements necessary to conduct research;
> - Identify the anticipated future spectrum requirements for at least the next 10 years; and

- Advise spectrum policy-makers on the value to the nation of accommodating scientific uses of the spectrum, recognizing the need to balance multiple communities.

The committee will comment on the spectrum use by the relevant scientific communities but will not make recommendations on the allocation of specific frequencies.

Appendix B

Biosketches of Committee Members

Marshall H. Cohen, ***Co-Chair,*** received his Ph.D. in physics from Ohio State University in 1952. He is a professor emeritus in the Astronomy Department at the California Institute of Technology (Caltech). Before going to Caltech, he was a professor of electrical engineering and then a professor of astronomy at Cornell University, spent 2 years as a professor of applied electrophysics at the University of California at San Diego, and then went to Caltech in 1968. Dr. Cohen has conducted radio astronomy research in solar physics and active galactic nuclei (AGN) and optical research on magnetic white dwarfs and on AGN. He was also involved with commissioning the Arecibo telescope and in developing Very Long Baseline Interferometry (VLBI) and the network that was set up to manage VLBI observations in the 1970s. Currently, he uses the Very Long Baseline Array to study the statistics of superluminal sources. Using the large telescopes at the Palomar and the W.M. Keck Observatories, he conducts polarization observations of the spectrum to study the relations among the different classes of objects and their evolution. Dr. Cohen has been very involved with activities of the National Research Council, having been a member of the Division on Engineering and Physical Sciences; the Commission on Physical Sciences, Mathematics, and Applications; the *Proceedings of the National Academy of Sciences* Editorial Board; the National Academy of Sciences (NAS) Class I Membership Committee; the United States National Committee for the International Union of Radio Science (USNC-URSI); chair of NAS Section 12: Astronomy; and chair of the 1980s Astronomy Survey Committee. He was also on panels of the 1970s and 1990s astronomy survey study.

Albin J. Gasiewski, ***Co-Chair,*** received his Ph.D. degree in electrical engineering and computer science from the Massachusetts Institute of Technology in 1989. Previously, he had received M.S. and B.S. degrees in electrical engineering and he received a B.S. degree in mathematics from Case Western Reserve University in 1983. From 1989 to 1997 Dr. Gasiewski was a faculty member within the School of Electrical and Computer Engineering at the Georgia Institute of Technology. As an associate professor there, he developed and taught courses on electromagnetics, remote sensing, instrumentation, and wave propagation theory. From 1997 through 2005 he worked at the U.S. National Oceanic and Atmospheric Administration's (NOAA's) Earth System Research Laboratory (ESRL) in Boulder, Colorado, where he was chief of the Microwave Systems Development Branch of the ESRL Physical Science Division. In 2006 he joined the Department of Electrical and Computer Engineering of the University of Colorado at Boulder, where he directs the NOAA-CU Center for Environmental Technology. His technical interests include passive and active remote sensing, radiative transfer theory and applications, electromagnetics, antennas and microwave circuits, electronic instrumentation, airborne sensors, meteorology, and oceanography. Dr. Gasiewski was the 2005-2006 president of the IEEE Geoscience and Remote Sensing Society and was the General Co-chair of IGARSS 2006 in Denver, Colorado. He is also a member of the International Union of Radio Science (URSI), where he currently serves as vice chair of the United States National Committee for the International Union of Radio Science (USNC-URSI) Commission F. He served on the U.S. National Research Council's Committee on Radio Frequencies (CORF) from 1989 to 1995 and on the USNC-URSI from 1996 to 1997.

Donald C. Backer is a professor of astronomy and chair of the Astronomy Department at the University of California, Berkeley. Professor Backer received a bachelor of engineering physics degree from Cornell University in 1966, a Master of Science degree in radio astronomy from Manchester University in 1968, and a Ph.D. in astronomy from Cornell University in 1971. He spent 2 years as a postdoctoral research assistant at the National Radio Astronomy Observatory (NRAO) in Charlottesville, Virginia, and 2 years as a National Research Council (NRC) Fellow at NASA's Goddard Space Flight Center. Since 1975 he has been at the University of California, Berkeley. His past duties have included serving as executive officer, and later chair, of the U.S. Very Long Baseline Interferometry (VLBI) network. More recently he has served on the board of the Berkeley-Illinois-Maryland Association and the visiting committees of the NRAO and the Haystack Observatory and is currently on the visiting committee of the Arecibo Observatory serving as chair. He chaired Commission J of the U.S. National Committee for the International Union of Radio Science from 1997 to 1999 and was an NRC ex officio member. Professor Backer's research interests have focused on pulsars and active galactic nuclei. One research effort involves the timing of an array of millisecond pulsars for use as celestial clocks. The long-term goal is setting limits on the gravitational wave background that may result from the coalescence of massive black holes in distant galaxies. His short-term goals include the investigation of small-scale turbulence in the interstellar plasma. He is involved with instrumentation for pulsar data acquisition at the Arecibo, Green Bank, Effelsberg, and Nançay Observatories. Another of Professor Backer's activities is focused on a deeper understanding of an enigmatic object in our galactic center, which may be the site of a massive black hole. VLBI observations at millimeter wavelengths are being pursued as well as proper motion measurements and also circular polarization. Professor Backer is a past member of the NRC's Committee on Radio Frequencies (former chair), the Atacama Large Millimeter Array Review Committee, and the 1980s Astronomy Survey Committee. He currently serves on the Committee on Astronomy and Astrophysics and the United States National Committee for the International Union of Radio Science.

Roberta Balstad (formerly Roberta Balstad Miller) is a senior research scientist at Columbia University and a senior fellow with the Center for International Earth Sciences Information Network (CIESIN). Dr. Balstad has published extensively on science policy, information technology and scientific research, remote sensing applications and policy, and the role of the social sciences in understanding global environmental change. She is the author of numerous articles and books, including *City and Hinterland: A Case Study of Urban Growth and Regional Development* (1979), and editor, with Harriet Zuckerman, of *Science Indicators: Implications for Research and Policy* (1980). Dr. Balstad received her Ph.D. from the University of Minnesota in 1974. She was a senior fellow at Oxford University in 1991 to 1992 and a guest scholar at the Woodrow Wilson International Center for Scholars in 1994. She is currently chair of the U.S. National Committee on Science and Technology Data (CODATA) and chaired the Priority Area Assessment Panel on Scientific Data and Information of the International Council for Science (ICSU). She is a member of the board of directors of the Open Geospatial Consortium and of the U.S. National Committee for the International Institute for Applied Systems Analysis. Before joining Columbia University, Dr. Balstad had been the director of the Division of Social and Economic Sciences at the National Science Foundation, the founder and first executive director of the Consortium of Social Science Associations (COSSA), and president/chief executive officer of CIESIN. In 1998, she led CIESIN's transition from Saginaw, Michigan, to become part of the Earth Institute at Columbia University, where she served as CIESIN's director through April 2006. Dr. Balstad has lectured widely, both in the United States and abroad. From 1992 to 1994, she was vice president of the International Social Science Council and has also served as chair of the NRC Steering Committee on Space Applications and Commercialization, of the NATO Advisory Panel on Advanced Scientific Workshops/Advanced

Research Institutes, of the American Association for the Advancement of Science's Committee on Science, Engineering and Public Policy, and of the Advisory Committee of the Luxembourg Income Study. She currently serves as chair of St. Antony's College Trust (Oxford University) in North America.

Steven W. Ellingson is an associate professor in the Bradley Department of Electrical and Computer Engineering at Virginia Polytechnic Institute and State University. Dr. Ellingson received his Ph.D. in electrical engineering from Ohio State University in 2000. Before going to Virginia Tech, he held research positions at Ohio State University, Raytheon, and Booz Allen Hamilton, Inc. Dr. Ellingson was previously a captain in the U.S. Army, on active duty between 1989 and 1993. His research interests are in the general areas of electromagnetics, applied signal processing, and instrumentation. He is specifically interested in direction finding, interference mitigation, wireless communications, radio astronomy, and the design of antennas and receivers. He has been working closely with the Long Wavelength Array. Dr. Ellingson is a member of the NRC's Committee on Radio Frequencies and is a senior member of the Institute for Electrical and Electronics Engineers.

Darrel Emerson was an assistant director of the National Radio Astronomy Observatory (NRAO), responsible for Arizona Operations, in Tucson, Arizona. His responsibilities included the operation of the NRAO 12-meter telescope at Kitt Peak, which undertakes fundamental astronomical research in the range 67-300 GHz. He is heavily involved in the Atacama Large Millimeter Array project. Dr. Emerson received his Ph.D. in radio astronomy in 1973 from the Cavendish Laboratory at the University of Cambridge, England. Before joining the NRAO, he worked for several years with the Max Planck Institute for Radio Astronomy 100-meter radio telescope at Effelsberg, near Bonn, Germany, and then with the Institute for Radio Astronomy in Millimeter-waves in Grenoble, France. His current research interests include spectral line studies of nearby normal galaxies and the development of millimeter-wave observational techniques.

Aaron S. Evans is an associate professor of astronomy at the University of Virginia and an associate astronomer at the National Radio Astronomy Observatory. He received his Ph.D. in astronomy from the Institute for Astronomy, University of Hawaii, in 1996. His current research primarily deals with observations of colliding galaxies and their associated phenomena (starbursts and active galactic nuclei). The study of these galaxies requires a multi-wavelength approach, which to date has included optical to mid-infrared imaging, as well as near-infrared and submillimeter spectroscopy. The observing facilities at which he carries out these programs are the Mauna Kea Observatories in Hawaii (University of Hawaii

2.2-meter, United Kingdom Infra-Red Telescope, James Clerk Maxwell Telescope, W.M. Keck Observatory); the Hubble Space Telescope; the Owens Valley Millimeter Array in California; the Steward Observatory 12-meter telescope at Kitt Peak, Arizona; and the IRAM 30-meter telescope in Spain. Dr. Evans received a NASA/American Society for Engineering Education Faculty Fellowship Award in 2002, and chaired the National Science Foundation's NRAO 5-Year Proposal Panel. He also served on the NRC's Committee to Review the Science Requirements for the Atacama Large Millimeter Array.

Joel T. Johnson is a professor of electrical and computer engineering in the Department of Electrical Engineering at the Ohio State University. He received his Ph.D. in 1996 from the Massachusetts Institute of Technology. Dr. Johnson's research interests include microwave remote sensing of geophysical media, both active and passive; the application of numerical techniques in electromagnetics to remote sensing problems; and the design of systems for radio frequency interference mitigation. He served from 2005 to 2009 as chair of the Frequency Allocations in Remote Sensing (FARS) Committee of the IEEE Geoscience and Remote Sensing Society; the FARS Committee's mission is to provide technical assessments, guidance, and recommendations regarding matters of frequency sharing and interference between remote sensing and other uses of the radio spectrum.

Paul Kolodzy is a private consultant with Kolodzy Consulting, LLC. He received his Ph.D. and M.S. in chemical engineering from Case Western Reserve University and his B.S. in chemical engineering from Purdue University. Prior to his work as a private consultant, he was the senior technology adviser and consultant to M2Z Networks. Before joining M2Z Networks he was the director of the Center for Wireless Network Security (WiNSeC) at the Stevens Institute of Technology. Prior to serving there, he was the senior spectrum policy adviser at the Federal Communications Commission (FCC) and the director of the Spectrum Policy Task Force charged with developing the next-generation spectrum policy. Dr. Kolodzy has also been a program manager in the Advanced Technology Office at the Defense Advanced Research Projects Agency (DARPA), managing research and development for communications programs developing generation-after-next capabilities. Before serving at DARPA, he had been the director of Signal Processing and Strategic Initiatives at Sanders (now BAE Systems), a premier electronic warfare company. Dr. Kolodzy got his start as the group leader and staff member at the Massachusetts Institute of Technology's Lincoln Laboratory, working on optical systems for laser radars, signal processing, and target recognition for acoustics, radio frequency (synthetic aperture radar), and optical signatures. Dr. Kolodzy has 20 years of experience in technology development for advanced communications, networking, electronic warfare, and spectrum policy for government, private–sector, and academic groups.

He participated in the NRC Computer Science and Telecommunications Board's Forum on Spectrum Management Policy Reform.

David B. Kunkee conducts microwave remote sensing research at Aerospace Corporation. This research is related to the development of the National Polar-orbiting Operational Environmental Satellite System, the Defense Meteorological Satellite Program, and NASA's Advanced Microwave Scanning Radiometer. Dr. Kunkee is active in radio science applications and is an amateur radio hobbyist. He is a member of Commission F of the International Union of Radio Science and is a member of the Institute of Electrical and Electronics Engineers' Geoscience and Remote Sensing Society, Antennas and Propagation Society, and Microwave Theory and Techniques Society. He received his Ph.D. in electrical engineering from the Georgia Institute of Technology in 1995.

Molly K. Macauley is a senior fellow and director of academic programs at Resources for the Future, Inc. (RFF). Dr. Macauley's research at RFF has included public finance, energy economics, the value of information, and economics and policy issues of outer space. She has been a visiting professor in the Department of Economics at the Johns Hopkins University. Dr. Macauley has testified before Congress on numerous occasions on topics including space commercialization, remote sensing, and legislative and regulatory space policy. She has served on many committees, including the congressionally mandated Economic Study of Space Solar Power for which she was chair. She currently serves on the Space Studies Board of the NRC, the Applied Sciences Advisory Group for NASA's Earth Sciences, and the Climate Working Group of NOAA's Science Advisory Board.

James M. Moran is a professor and senior radio astronomer at the Harvard-Smithsonian Center for Astrophysics and chair of the Department of Astronomy at Harvard University. He has made fundamental and far-ranging contributions to astronomy through his key developments of radio spectroscopy combined with interferometry. He has used these techniques to study cosmic masers and has obtained, among other important results, direct and definitive evidence for the existence of a supermassive black hole. Dr. Moran observes molecular masers to study the dynamics of gas surrounding putative black holes in nearby galaxies. These masers can be tracked precisely in position and velocity with intercontinental arrays of radio telescopes operating as very long baseline interferometers. With the high angular resolution provided by these interferometers, he is able to measure the orbital characteristics of the gas as well as the mass and location of the black hole. Dr. Moran was principal investigator of the Sub-millimeter Array, an eight-element linked interferometric array, built near the summit of Mauna Kea in Hawaii and used to study planetary atmospheres, star formation, quasars, dust and gas distribution in nearby galaxies, and spectral lines from highly redshifted galaxies. Dr. Moran served on the U.S. National Committee for the International Astronomical Union (member, January 2000 through December 2002), the Astronomy and Astrophysics Survey Committee (member, August 1998 through June 2002) and its Panel on Radio and Sub-millimeter-wave Astronomy (vice chair, November 1998 through December 2001), the U.S. National Committee for the International Union of Radio Science (ex officio member, January 1991 through December 1993), and CORF's Subcommittee on Radio Astronomy (member, July 1984 through June 1987). Dr. Moran is a member of the National Academy of Sciences.

Lee G. Mundy is a professor and chair of the Department of Astronomy at the University of Maryland at College Park. He received his Ph.D. in astronomy in 1984 from the University of Texas at Austin. Dr. Mundy studies the dense interstellar medium, star formation, and the initial stages of planet formation utilizing observations at centimeter through near-infrared wavelengths and radiative transfer modeling tools. The observations are mainly acquired with the Very Large Array and Berkeley Illinois Maryland Association/Combined Array for Research in Millimeter-wave Astronomy, and through a Space Infrared Telescope Facility legacy project that is mapping five major molecular clouds and more than 100 compact cores. Dr. Mundy is also collaborating with the NASA Goddard Space Flight Center in studies of a number of mission concepts for submillimeter through near-infrared wavelength space interferometers. Dr. Mundy has published extensively.

Timothy J. Pearson is a senior research associate at the California Institute of Technology. He received his Ph.D. from the University of Cambridge in 1977, after which he held a postdoctoral position at Caltech and has been at Caltech since then. Dr. Pearson's research interests include the statistics of radio sources, and radio interferometry and its application to observations of active galactic nuclei and the cosmic microwave background radiation. He uses radio telescopes at Cambridge, the Owens Valley Radio Observatory, the National Radio Astronomy Observatory, and the Cosmic Background Imager in Chile. Currently he is an associate editor for the *Monthly Notices of the Royal Astronomical Society.*

Christopher S. Ruf is a professor in the Department of Atmospheric, Oceanic, and Space Sciences and in the Department of Electrical Engineering and Computer Sciences at the University of Michigan. He is also director of the Space Physics Research Laboratory. He received his Ph.D. in electrical and computer engineering from the University of Massachusetts at Amherst. Dr. Ruf works in microwave radiometry, an important area of remote sensing and radio frequency protection issues. His research interests include Earth environmental remote sensing, synthetic thinned aperture radiometry, the mitigation of radio frequency interference, self-contained end-to-end radiometer calibration systems, the use of stationary statistical properties of upwelling radiances to constrain absolute accuracy and the long-term stability of satellite measurements, and the profiling of the lower, middle, and upper atmosphere using multispectral, multisensor, and climatological databases. Before assuming his position at the University of Michigan, Dr. Ruf had been an instrument scientist for the NASA TOPEX and Jason-I microwave radiometers, and he is currently a science team member for the NASA Juno, Aquarius, and Global Precipitation Measurement (GPM) Microwave Imager microwave radiometers. Dr. Ruf has received numerous awards, including the International Geoscience and Remote Sensing Symposium Prize Paper Award. He is the editor in chief of the *IEEE Transactions on Geoscience and Remote Sensing*, a member of URSI Commission F, and a past member of the NRC's Committee on Radio Frequencies.

Frederick S. Solheim is the president of Radiometrics Corporation, where he develops ground-based microwave radiometers for atmospheric and terrestrial remote sensing. Dr. Solheim was heavily involved with the development of the patented frequency-agile design that allows flexibility for a variety of atmospheric remote sensing applications used in the company's radiometers. His research interests include microwave radiometry and radiosonding for profiles of temperature, water vapor, and cloud liquid. Dr. Solheim also conducts research in signal propagation. Previously he worked with the University Corporation for Atmospheric Research in Boulder, Colorado.

David H. Staelin is a professor of electrical engineering in the Department of Electrical Engineering and Computer Science (EECS) at the Massachusetts Institute of Technology (MIT). He has been a member of the EECS faculty and the Research Laboratory of Electronics since 1965. He also was the assistant director, MIT Lincoln Laboratory (1990-2001); co-founder, MIT Venture Mentoring Service (2000); chair, MIT's EECS Graduate Area in Electronics, Computers, and Systems (1976-1990); and a faculty member of MIT's Leaders for Manufacturing Program (1985-1998). He was a director of Environmental Research and Technology, Inc. (1969-1978) and a co-founder and chair of PictureTel Corp. (1984-1987). His research interests include remote sensing, wireless communications, signal processing, estimation, environmental sensing, microwave atmospheric sounding, and meteorological satellites. Dr. Staelin was a member of the President's Information Technology Advisory Committee (2003-2005), chair of the NRC's Committee on Radio Frequencies (1983-1986), and a member of several NASA committees and working groups, including the Space Applications Advisory Committee, the Advanced Microwave Sounder Working Group, the Geostationary Platform—Earth Science Steering Committee; and the Tropical Rainfall Measuring Mission Science Steering Group. He was the principal investigator for the NASA Nimbus-E Microwave Spectrometer (launched in 1972 on Nimbus-5) and the Scanning Microwave Spectrometer (launched in 1975 on Nimbus-6). He was a co-investigator of the Scanning Multichannel Microwave Spectrometer (launched in 1977 on Nimbus 7) and the Voyager Planetary Radio Astronomy Experiment (launched in 1977 on Voyagers 1 and 2). Additionally, he is a member of the NASA Atmospheric Infrared Sounder team (Aqua launched in 2002), the NPOESS Preparatory Project Science Team (to be launched in about 2010), the NOAA International Program Office Sounder Operational Algorithm Team, and the NASA Precipitation Mapping Mission Science Team. Dr. Staelin is a fellow of the IEEE and American Association for the Advancement of Science and received the 1996 Distinguished Achievement Award from the IEEE Geoscience and Remote Sensing Society.

Alan B. Tanner is an engineer at the NASA Jet Propulsion Laboratory. He received his Ph.D. in electrical engineering in 1989 from the University of Massachusetts at Amherst. His research interests include propagation, aperture synthesis, radiometers, and sounding. Dr. Tanner is involved in GeoSTAR, a microwave sounder intended for geosynchronous orbit.

Appendix C

Density of Interferers Equation

The power received at a radiometer due to the emission of P_T watts from a particular interferer at distance R can be predicted using the Friis formula:

$$P_R = \frac{P_T G_T}{4\pi R^2} A_{eff} e^{-\tau} \tag{C.1}$$

where P_R is the power received at the radiometer in watts, G_T is the transmitter antenna gain in the direction of the radiometer (dimensionless), A_{eff} is the effective aperture of the receive antenna (square meters), and $e^{-\tau}$ describes the attenuation of the transmitted power by atmospheric gases, clouds, and rain along the path from the transmitter to the receiver. The product $P_T G_T$ when using the maximum of the transmitter antenna gain is also referred to as the equivalent isotropic radiated power (EIRP) of a source. For multiple uncorrelated radio frequency interference (RFI) sources within a radiometer footprint, the EIRP of the interference is usually approximated as the sum of that of all the individual sources.

The received power P_R produces a brightness-temperature perturbation of

$$\delta T = \frac{P_R}{kB}(\mathrm{K}) \tag{C.2}$$

where k is Boltzmann's constant (1.38×10^{-23} W – Hz^{-1}K^{-1}) and B is the radiometer bandwidth (Hz). Combining Equations C.1 and C.2 and using the property that the radiometer beamwidth (and hence footprint size) is related to the antenna size (and hence to the square root of the effective aperture area), the density (in W/m^2) of the EIRP within the radiometer field of view can be related to the maximum tolerable brightness perturbation:

$$\frac{P_T G_T}{A} = \delta T \frac{kBe^{\tau}}{\lambda^2}\left(\frac{64}{\pi}\right)\left(\mathrm{Wm}^{-2}\right) \tag{C.3}$$

where A is the radiometer footprint area (m^2); Equation C.3 shows that it is the density of EIRP per area (computed over the radiometer footprint) that determines the interference to the radiometer. EIRP limits on individual transmitters must be combined with information on the expected number of transmitters within a specific area in order to predict or interpret observed interference levels δT.

As an example, a 6.9 GHz Advanced Microwave Scanning Radiometer-Earth (AMSR-E) observation (2,500 km^2 footprint area) with a bandwidth of 350 MHz will experience a brightness increase of 1 K if even a single interferer having a 130 milliwatt EIRP (in the direction of the radiometer antenna) is included in the footprint area. That such low radiated interference powers can perturb observed brightness temperatures demonstrates the high sensitivity of Earth Exploration-Satellite Service observations to interference. The fact that multiple interference sources may reside within any radiometer footprint substantially exacerbates the problem. The impact of a specific interference level on a particular geophysical measurement depends on the sensitivity of the measurement to changes in brightness temperature, as discussed in §2.2 in Chapter 2 of this report. The accuracy achieved in current radiometer systems typically makes even small changes in brightness caused by radio frequency interference to have a significant impact on measured products.

Appendix D

Analysis of Out-of-Band Emission Impacts to the EESS from §27.53 of the FCC Rules

The general analysis presented here is applicable both to Earth Exploration-Satellite Service (EESS) radiometers and to Radio Astronomy Service (RAS) radiometers. EESS and RAS radiometers are governed by the same technical principles, and for both, a source of radio frequency interference (RFI) operating in compliance with Federal Communications Commission (FCC) rules can deleteriously affect a radiometer operating in a primary protected band, as described below.

Parameters of two Federal Aviation Administration (FAA) Air Route Surveillance Radars (ARSRs) are given in Table D.1, from Piepmeier and Pellerano.[1]

According to FCC rules, 47 C.F.R., §27.53, Part (j) (page 387) out-of-band (OOB) emission limits for the class of radars specified in Table D.1 are as follows:

> For operations in the unpaired 1390-1392 MHz band and the paired 1392-1395 MHz and 1432-1435 MHz bands, the power of any emission outside the licensee's frequency band(s) of operation shall be attenuated below the transmitter power (P, in Watts) by at least (43 + 10 log (P)) dB.[2]

Note that the "log" function refers here to the base 10 logarithm. In order to determine if the specified attenuation is achieved, measurement of the out-of-band

[1] J.R. Piepmeier and F. A. Pellerano, "Mitigation of Terrestrial Radar Interference in L-Band Spaceborne Microwave Radiometers," in *Proceedings of the 2006 International Geoscience and Remote Sensing Symposium (IGARSS)*, Denver, Colorado, 2006, pp. 2292-2296, DOI 10.1109/IGARSS.2006.593.

[2] See 47 C.F.R., §27.53, Part (j), available at http://ecfr.gpoaccess.gov/; accessed January 15, 2010.

TABLE D.1 Parameters of Two Federal Aviation Administration Air Route Surveillance Radars

Name	Frequency (MHz)	Peak Power (kW)	Antenna Gain (dBi)	Azimuth Beamwidth (degrees)	Scan Rate (rpm)	Pulse Width (μsec)	Pulse Repetition Frequency (PRF) (Hz)
ARSR-3	1250-1350	5000	34	1.25	5	2	310-365
ARSR-4	1215-1400	60	35	1.4	5	9/60	216/72

SOURCE: J.R. Piepmeier and F.A. Pellerano, "Mitigation of Terrestrial Radar Interference in L-Band Spaceborne Microwave Radiometers," in *Proceedings of the 2006 International Geoscience and Remote Sensing Symposium (IGARSS)*, Denver, Colorado, 2006, pp. 2292-2296, DOI 10.1109/IGARSS.2006.593.

radiated power in a 1 MHz bandwidth is specified by the FCC. It is therefore possible to radiate larger out-of-band total powers in bandwidths larger than 1 MHz. These regulations specify for the ARSR-4 radar, for example, that the allowed OOB emission is $P_{t\_OOB} = 10 \log(6 \cdot 10^4\ W) - (43 + 10 \log(6 \cdot 10^4\ W)) = -43$ dBW (peak) in a 1 MHz bandwidth.

The Friis formula specifies the power received by an EESS radiometer from a transmitting source:

$$P_{RFI} = \frac{1}{L_{FDR}} P_t G_t(\theta_r, \phi_r) G_r(\vartheta_t, \phi_t) \left(\frac{\lambda}{4\pi R}\right)^2$$

where L_{FDR} is the frequency dependent rejection (FDR) factor, P_t is the transmit power of the radar, $G_t(\theta_t, \phi_t)$ is the gain of the radar transmitting antenna in the direction of the radiometer, $G_r(\theta_r, \phi_r)$ is the gain of the radiometer antenna in the direction of the radar, λ is the wavelength of the radar frequency, and R is the range between the radar and the radiometer. Using this equation to test the permissible spurious and OOB power levels according to §27.53, set $L_{FDR} = 1.0$, since it is assumed that the OOB emissions occur within the radiometer bandwidth.

Assume that $G_t(\theta_t, \phi_t) \approx 20$ dBi (~ –15 dB from maximum gain due to elevation differences in the line-of-sight [LOS] to the space-based radiometer), $G_r(\theta_r, \phi_r)$ of the Aquarius (or similar) radiometer is ~25 dB, $\lambda = 0.21$ m, and $R \approx 1 \times 10^6$ m LOS from a low Earth orbit (LEO) of altitude ~700 km. These values result in: $P_{RFI\_OOB} \sim (-43\text{ dBW}) + (20\text{ dBi}) + (25\text{ dBi}) + (-155.5\text{ dB}) = -153.5$ dBW for an ARSR-4 radar system. This is a peak power level whose impact would be reduced when integrated over a longer integration period; the most conservative (i.e., shortest) relevant ratio of the radar pulse width to the radiometer integration time is $\sim(6 \times 10^{-5})/(1 \times 10^{-3}) = 6 \times 10^{-2} \approx -12.2$ dB. However, the OOB received power is increased by the fact that the EESS radiometer passband is ~27 MHz, compared to the 1 MHz bandwidth specified in §27.53(a)4; the case of OOB emissions at the permitted level throughout the entire 27 MHz bandwidth adds 14.3 dB to P_{t_OOB}.

Therefore, for a single integration time of 1 ms, the spurious power received by the EESS radiometer from a single radar may be ~ –153.5 – 12.2 + 14.3 = –151.4 dBW. In contrast, the single sample sensitivity of an L-band EESS radiometer with a 1 ms integration time can be derived using similar parameters:

$$T_{REC} = 290 \cdot \left(10^{\frac{NF}{10}} - 1 \right) \approx 20\text{K}$$

$$T_{SYS} = T_{REC} + T_{ANT} \approx 100\text{K}$$

for H-polarization over the ocean. Therefore the sensitivity is

$$\Delta T_{RMS} = \frac{100}{\sqrt{\tau_{\text{int}} BW}} = \frac{100}{\sqrt{\left(1 \cdot 10^{-3}\right)\left(27 \cdot 10^{6}\right)}} \approx 0.61\text{K}.$$

The minimum detectable change in power of the EESS radiometer with a factor of 10 safety margin for a single sample is as follows:

$$k\Delta TB = 1.38 \cdot 10^{-23} \left(\text{J} \cdot \text{K}^{-1} \right) \cdot 27 \cdot 10^{6}\,\text{s}^{-1} \cdot 0.06\text{K} \approx -166.5\text{dBW}.$$

In this scenario there is safety margin of ~10 in the impact from a single radar. This means that the OOB emission requirements are inadequate to protect EESS, and they do not even closely meet the expectations of International Telecommunication Union-Radio (ITU-R) RS.1029 that interfering signal levels should be below –171 dBW within a 27 MHz bandwidth at 1.4 GHz by roughly 5 dB. Note that this analysis is for a single radar within the footprint of the radiometer. More than one radar in the radiometer field of view results in further reduction of the safety factor and errors in the data that are virtually impossible to detect without auxiliary information. Unfortunately, it appears that limits of the adjacent signal rejection of the EESS radiometer (due to filter limitations) result in additional contamination of the EESS radiometer field as detailed in Piepmeier and Pellerano.[3]

[3] J.R. Piepmeier and F.A. Pellerano, "Mitigation of Terrestrial Radar Interference in L-Band Spaceborne Microwave Radiometers," in *Proceedings of the 2006 International Geoscience and Remote Sensing Symposium (IGARSS)*, Denver, Colorado, 2006, pp. 2292-2296, DOI 10.1109/IGARSS.2006.593.

In EESS radiometer systems, RFI levels are cumulative. Therefore, impacts from adjacent signals described by Piepmeier and Pellerano,[4] coupled with the additional impacts from spurious and OOB emissions in the above analysis, suggest that a single ARSR-type radar operating in full compliance with FCC §27.53 can impact the operation of EESS radiometers operating in the 1.4 GHz protected region.

This analysis has assumed that the OOB emissions from the radar are at the maximum allowable level (–43 dBW/MHz) throughout the entire 27 MHz radiometer bandwidth; that the radar transmits its peak power over its pulse width, which lies entirely within the radiometer integration period; and that the antennas are oriented so that 15 dB below maximum antenna coupling occurs.

[4]J.R. Piepmeier and F.A. Pellerano, "Mitigation of Terrestrial Radar Interference in L-Band Spaceborne Microwave Radiometers," in *Proceedings of the 2006 International Geoscience and Remote Sensing Symposium (IGARSS)*, Denver, Colorado, 2006, pp. 2292-2296, DOI 10.1109/IGARSS.2006.593.

Appendix E

Descriptions of Earth Exploration-Satellite Service Parameters Related to Table 2.1

AIR TEMPERATURE PROFILES

Global air temperature profiles are critical to numerical weather predictions (NPWs) because temperature is inversely related to air density and therefore to the differential gravitational forces on air that help drive local and global winds. Temperature also serves as a tracer of atmospheric motion. Although satelliteborne infrared imaging spectrometers can measure these profiles, clouds often introduce significant errors, particularly in the lower troposphere, in certain polar seasons, and in "baroclinic" regions, that commonly exert a disproportionate influence on future weather events. Current operational weather satellites combine both microwave and infrared spectrometer data to take advantage of the relative strengths of each; this sensor combination probably makes the single most important contribution of weather satellite data to the dramatic improvements achieved in providing useful global weather predictions up to a week in advance. Although the 50-60 GHz oxygen absorption bands provide most such data, they are generally supplemented by other bands that help correct the results for precipitation, humidity, and surface effects, discussed below in this appendix. In addition, it has been found that the original operational temperature sounding microwave instruments (microwave sounding units, or MSUs) can be calibrated across satellites to yield a very sensitive indicator of global warming in the middle troposphere with accuracies on the order of 0.1 K per decade. These observations are being continued using successor instruments such as advanced MSUs (AMSUs). Systematic radio frequency interference (RFI) at levels too low to be otherwise detected could, in principle, introduce errors in such measurements.

PRECIPITATION RATE

Observations of global precipitation are important to both weather forecasts and climate studies. They are particularly useful in monitoring severe storms such as hurricanes and damaging fronts. Precipitation is important not only to safety, agriculture, and commerce, but also to hydrology and predictions of floods, soil moisture, and sea surface salinity (SSS). Since the locations of convective precipitating cells cannot be predicted well, and because they sometimes reside under higher cloud shields such as those obscuring hurricanes and other severe storms, only microwave sensors can reveal their intensities and locations. Precipitation is generally observed using the same sensors as those used for water vapor, which include (1) window-channel sensors at frequencies such as 18.7, 22, 23.8, 31.4, 37, and 89 GHz that observe raindrop emission against colder backgrounds such as ocean and low-emissivity soil (e.g., Special Sensor Microwave/Imager [SSM/I], Special Sensor Microwave Imager/Sounder [SSMI/S], Advanced Microwave Scanning Radiometer-Earth [AMSR-E]), and (2) the opaque water vapor resonance 176-191 GHz in combination with lower frequencies such as 89, 150, and 164-168 GHz; glaciated cell tops are particularly visible and sensitive to convective strength. In addition, the opaque oxygen bands 50-56 GHz are useful because they are sensitive to ice particle size distributions and therefore to the heavier precipitation rates (e.g., AMSU, SSMI/S).

SEA SURFACE SALINITY

Sea surface salinity is a critical missing parameter that scientists need in order to meet climate research goals. Measuring global SSS over time will contribute to scientists' understanding of change in the global Earth system and how the system responds to natural and human-induced change. Global measurements of SSS can be achieved to ~0.2 practical salinity units using space-based passive microwave radiometry at 1.4 GHz and radar scatterometry at 1.26 GHz.[1] These measurements can provide significant new information about how global precipitation, evaporation, and the water cycle are changing. Global SSS variability provides key insight regarding freshwater flow into and out of the ocean associated with precipitation, evaporation, ice melting, and river runoff. Global SSS measurements will also provide important background about how climate variation induces changes in global ocean circulation. The combination of global SSS and sea surface temperature

[1] See http://aquarius.nasa.gov/science.php; accessed on January 15, 2010.

(SST) measurements can be used to determine seawater density, which regulates ocean circulation and the formation of water masses.

SEA SURFACE TEMPERATURE

Global all-weather sea surface temperature data are critical for NWP and climate research. SST measurements are important for understanding heat exchange and coupling between the ocean and atmosphere, and SST data are required by operational ocean analyses in order to properly constrain upper-ocean circulation and thermal structure.[2] SST measurements in clear air can be obtained using electro-optical (traditional) instruments; however, clouds prevent these measurements, so passive microwave measurements within the 5 to 6 GHz region are critical for obtaining coverage in areas that are seasonally cloud-covered. For example, areas in the U.S. Exclusive Economic Zone off the coast of Washington and Oregon are not imaged with traditional satellite SST sensors for weeks at a time owing to persistent stratus cloud cover, necessitating an all-weather solution. The standard SST measurement uncertainty for space-based SST measurements is 0.5 K at 50 km (passive microwave, all-weather capability). To achieve this standard for microwave measurements, interfering signal power within a (typical) receiving bandwidth of 350 MHz (e.g., AMSR-E) must be below approximately –126 dBm[3] using a factor of 10 power margin. For reference, this power level is effectively 3 dB higher than recommended levels from International Telecommunication Union-Radio (ITU-R) RS.1029, but still far below the level of interference that would be considered acceptable for nearly all other communication and signal systems. Space-based SST measurements near 6 GHz near land are impacted primarily by land-based emitters operating in the fixed service within full compliance of their regulations. Less pervasive RFI impacts are encountered from shipboard radar. For SST measurements using 10.7 GHz such as TRMM's TMI and AMSR-E, substantive RFI is incurred from geostationary transmitters operating immediately adjacent to the upper edge of the 10.7 GHz EESS band segment as depicted in Figure 2.16 in Chapter 2 in this report.

SOIL MOISTURE

Global, high-quality soil moisture measurements are expected to advance weather forecasting and Earth hydrology studies significantly. A proposed NASA

---

[2]C.J. Donlon, P.J. Minnett, C. Gentmann, T.J. Nightingale, I.J. Barton, B. Ward, and M.J. Murray, "Toward Improved Validation of Satellite Sea Surface Skin Temperature Measurements for Climate Research" *Journal of Climate,* 15: 353-369 (February 2002).

[3]With a factor of 0.5 sensitivity, this value is $(1.38 \times 10^{-23}\ \mathrm{J} \cdot \mathrm{K}^{-1})(0.05\ \mathrm{K})(350 \times 10^{6}\ \mathrm{Hz}) = 2.42 \times 10^{-16}\ \mathrm{W}$.

mission, Soil Moisture Active Passive (SMAP), would provide measurements of soil moisture using 1.4 GHz passive microwave radiometry and 1.26 GHz radar scatterometer to measure soil moisture to approximately 4 percent uncertainty with about 1.5 kg/m^2 surface vegetation water content. Soil moisture measurements at higher frequencies, such as those planned for the National Polar-orbiting Operational Environmental Satellite System (NPOESS) near 6 and 10 GHz, will also provide additional data refresh reducing data latency and measurements capable of producing soil moisture estimates to approximately 8 percent uncertainty at 50 km horizontal spatial resolution.

A National Centers for Environmental Prediction (NCEP) Scientific Assessment has determined that the NCEP Eta model requires soil moisture to properly calculate the energy fluxes at the surface. To support the model, the U.S. Department of Commerce requires measurements at the surface with a horizontal resolution of 50 km, mapping uncertainty of 3 km, and measurement accuracy of approximately 10 cm of water per 1 meter column of soil.[4]

SEA SURFACE WIND VECTOR

Space-based remote sensing of sea surface winds (SSWs) depends on precision measurements of polarimetric upwelling microwave emissions from the ocean surface at 10.7-37.0 GHz. High-quality SST measurements based on 6 GHz region brightness temperatures are also required to produce the best SSW direction product. Global SSW data are critical for high-quality NWP forecasts, developing tropical cyclone warnings, aircraft and ship operations, ship routing, and other civil and military operations. SSW data constitute one of the most important parameters (Environmental Data Records, or EDRs) in operational meteorological remote sensing. The accuracy of the wind vector products obtained from WindSat retrievals to date has reached or exceeded that available from active scatterometer systems such as QuikScat at moderate to high wind speeds, and the ability of microwave radiometers to simultaneously measure atmospheric and sea temperature properties motivates attempts to improve the accuracy of the radiometer products further.

SEA ICE

One of the first applications of space-based passive microwave imagery was for monitoring sea ice characteristics. The Electrical Scanning Microwave Radiometer (ESMR) data set provides the earliest all-weather, all-season imagery of polar

[4]NPOESS Integrated Operational Requirements Document (IORD), Acquisition Decision Memorandum 01-xxx, March 16, 2001.

sea ice. Some satellite data of sea ice in the visible and infrared wavelengths were available in the late 1960s and early 1970s (before the introduction of space-based passive microwave observations), but since the polar regions are either dark or cloud-covered for much of the year, the generation of consistent, long-term data records from visible and infrared sensing was not practical.

Passive microwave data introduced a major advance in the usefulness of satellite sea ice imaging. The value of passive microwave data for sea ice studies derives from the large contrast in microwave emissivities between sea ice and open water. At 19.35 GHz, open water has an emissivity of approximately 0.44, whereas various sea ice types have emissivities ranging from approximately 0.8 to 0.97. The resulting contrast in microwave brightness temperatures allows accurate estimates of sea ice concentrations (percentages of ocean area covered by sea ice) and hence the identification of sea ice distributions throughout the region of observation, as well as temporal variations of these distributions throughout the time period of observation.[5]

WATER VAPOR PROFILES

Global water vapor profiles are essential to the NWP of rainfall and drought, and they help constrain such predictions in general. As in the case of temperature profile measurements, combined microwave and infrared spectral data can yield near-all-weather global performance despite most clouds. Two different types of microwave observations are used, those in transparent bands within which the water vapor absorption stands out against the colder ocean background (ocean partially reflects the extremely cold cosmic background radiation), or against that of cold low-emissivity land. No profile information is usually retrieved, only an estimate of the column-integrated abundance. The frequencies most often used for this purpose include 18.7, 22, 23.8, 31.4, 37, and 89 GHz. To improve retrieval accuracies, these channels are often dual-polarized (horizontal and vertical) and scanned at a constant angle of incidence (e.g., SSM/I, SSM/IS, and AMSR-E). In addition, the opaque water vapor resonance near 183 GHz is often used in combination with some of the lower frequencies; these frequencies generally include 89, 150, 164-168, and 176-191 GHz, but must be used in combination with temperature profile information to yield the most accurate results (e.g., AMSU, SSM/IS). Instruments retrieving water vapor profiles are generally used to retrieve other parameters simultaneously, such as cloud water content, precipitation rate, ice and snow information, and so on.

[5]T.T. Wilheit, *Nimbus-5 User's Guide,* NASA/Goddard Space Flight Center, Greenbelt, Maryland, pp. 59-105.

CLOUD WATER

The ability of microwave radiometers to measure water vapor and cloud water directly is a significant capability, provided by no other remote sensing system. Radars measure cloud reflectivity, which has a strong dependence on water droplet size. Uncertainty in the cloud droplet size distribution makes radar measurements of cloud water inaccurate. Because liquid water is a strong absorber (and hence emitter) of microwave energy, the volume of cloud water can be more accurately measured with microwave radiometers. The microwave technique is also far more accurate than infrared or optical methods owing to the high reflectance, or albedo, of clouds at these wavelengths.

NUMERICAL WEATHER PREDICTION

In general NWP models such as the European Centre for Medium-Range Weather Forecasts (ECMWF), Navy NoGAPS use a full range of passive microwave data: 19.35, 22.235, 23.6-24.0, 31.3-31.8, 37, 50.3-57.3, 85.5, 89, 150, 176-190 GHz operationally. Although space-based global microwave observations have their largest impact where other data sources are sparse, significant positive impact is also identified in data-rich areas. It is estimated that in the Southern Hemisphere, microwave observations provide 60 to 70 percent of the impact of all satellite data in the ECMWF model. The total proportional impact, that is, the relative reduction if a particular band is lost, is over 50 percent for the band 54-57 GHz alone. Similarly, a loss of 24 GHz data would represent 30 percent of the total impact from microwave measurements. Note that these estimates assume that all other data remain intact, so the loss of more than one channel is more serious than a linear combination of losses would suggest. Other bands at a similar level of importance are 31.3-31.8, 57-59, 89, and 183.31 GHz.[6]

TRACE GASES

Although the protective stratospheric ozone layer is indeed recovering, the need for passive millimeter- and submillimeter-wave monitoring continues today. The ability to monitor the individual abundance, spatial distribution, and temporal trend of each of the trace species that contributes to the depletion process allows the efficacy of the Montreal Protocol to be directly verified. More importantly, recent, large-scale changes in the stratospheric makeup suggest that the rate of recovery of the ozone layer may be slowing.

---

[6]S. English, *The Value of Passive Microwave Satellite Observations to NWP*, Forecasting Research Technical Report No. 484, Exeter, U.K.: Met Office, 2006.

Millimeter- and submillimeter-wave frequencies distributed from approximately 183 to 916 GHz are ideally suited for observing ice clouds.[7] These high frequencies are necessary in order for scattering to be the dominant interaction mechanism. The wide range of frequencies accommodates the large dynamic range of ice water path that occurs in nature and is incorporated in the Submillimeter Infrared Radiometer Ice Cloud Experiment (SIRICE) mission that is currently in pre-Phase A development at NASA. See Table 2.4 in Chapter 2 of this report.

[7] A.J. Gasiewski, "Numerical Sensitivity Analysis of Passive EHF and SMMW Channels to Tropospheric Water Vapor, Clouds, and Precipitation," *IEEE Transactions on Geoscience and Remote Sensing,* 30: 859-870 (1992); K.F. Evans and G.L. Stephens, "Microwave Radiative Transfer Through Clouds Composed of Realistically Shaped Ice Crystals. Part II: Remote Sensing of Ice Clouds," *Journal of the Atmospheric Sciences,* 52: 2058-2072 (1995); K.F. Evans, S.J. Walter, A.J. Heymsfield, and M.N. Deeter, "Modeling of Submillimeter Passive Remote Sensing of Cirrus Clouds," *Journal of Applied Meteorology,* Vol. 37 (1998); K.F. Evans, A.H. Evans, I.G. Nolt, and B.T. Marshall, "The Prospect for Remote Sensing of Cirrus Clouds with a Submillimeter-wave Spectrometer," *Journal of Applied Meteorology,* 38: 514-525 (1999).

Appendix F

Acronyms and Abbreviations

ACT	Atacama Cosmology Telescope
AGN	active galactic nucleus
ALMA	Atacama Large Millimeter Array
AMSR-E	Advanced Microwave Scanning Radiometer-Earth
AMSU	Advanced Microwave Sounding Unit
ARM	Atmospheric Radiation Monitoring
ARSR	Air Route Surveillance Radars
ATA	The Allen Telescope Array
ATMS	Advanced Technology Microwave Sounder
AUI	Associated Universities, Inc.
AVHRR	Advanced Very High Resolution Radiometer
AWS	Advanced Wireless Services
BOINC	Berkeley Open Infrastructure for Network Computing
BRS/EBS	Broadband Radio Service and Educational Broadband Service
CAPE	connective available potential energy
CCAT	Cornell Caltech Atacama Telescope
CFC	chlorofluorocarbon
CMB	cosmic microwave background
CME	coronal mass ejection
CMIS	Compact Microscope Imaging System
COBE	Cosmic Background Explorer

CORF	Committee on Radio Frequencies
CR	cognitive radio
DDA	direct data assimilation
DFS	dynamic frequency selection
DIA	Denver International Airport
DMSP	Defense Meteorological Satellite Program
DOD	Department of Defense
ECV	essential climate variable
EDR	Environmental Data Record
EESS	Earth Exploration-Satellite Service
EIRP	equivalent isotropically radiated power
EOS	Earth Observing System
ESMR	Electrical Scanning Microwave Radiometer
ESTAR	Electronically-Scanned Thinned Array Radiometer
EVLA	Expanded Very Large Array
FASR	Frequency Agile Solar Radiotelescope
FCC	Federal Communications Commission
FS	Fixed Service
GBT	Green Bank Telescope
GCOM-W	Global Change Observation Mission-Water
GCOS	Global Climate Observing System
GEO	geostationary orbit (satellite); Group on Earth Observations
GEOSS	Global Earth Observation System of Systems
GFSC	Goddard Space Flight Center
GMI	Global Precipitation Measurement-Microwave Imager
GOES	Geostationary Operational Environmental Satellites
GPM	Global Precipitation Measurement (mission)
GPS	Global Positioning System
GR	Einstein's theory of general relativity
HCFC	hydrochlorofluorocarbon
HFC	hydrofluorocarbon
HOMER	Hybrid Optimization Model for Electric Renewables
HSB	Humidity Sounder for Brazil
IPCC	Intergovernmental Panel on Climate Change
ITU	International Telecommunication Union

ITU-R	ITU Radiocommunication Sector
ITU-RR	ITU Radio Regulations
IUCAF	Scientific Committee on Frequency Allocations for Radio Astronomy and Space Science
IWP	ice water path
JMR	JASON Microwave Radiometer
LEO	low Earth orbit
LI	lifted index
LIGO	Laser Interferometer Gravitational Wave Observatory
LISA	Laser Interferometer Space Antenna
LMS	Land Mobile Service
LMSS	Land Mobile Satellite Service
LMT	Large Millimeter Telescope
LWA	Long Wavelength Array
MEMS	microelectromechanical systems
MHS	Microwave Humidity Sounder
MIRAS	Microwave Imaging Radiometer using Aperture Synthesis
MIS	Microwave Imager/Sounder
MLS	Microwave Limb Sounder
MODIS	Moderate Resolution Imaging Spectroradiometer
MSU	microwave sounding unit
MWA	Murchison Widefield Array
NAIC	National Astronomy and Ionosphere Center
NASA	National Aeronautics and Space Administration
NEMS	Nimbus-E Microwave Spectrometer
NOAA	National Oceanic and Atmospheric Administration
NPOESS	National Polar-orbiting Operational Environmental Satellite System
NPRM	notice of public rule making
NRAO	National Radio Astronomy Observatory
NRC	National Research Council
NRQZ	National Radio Quiet Zone
NSF	National Science Foundation
NTIA	National Telecommunications and Information Administration
NWP	numerical weather prediction
OOB	out-of-band (emissions)

PATH	Precipitation and All-Weather Temperature and Humidity
PECAD/CADRE	Production Estimate and Crop Assessment Division's Crop Condition Data Retrieval and Evaluation
pfd	power flux density (usually measured in Wm^{-2})
PSR	Polarimetric Scanning Radiometer
PSU	practical salinity unit
PWV	precipitable water vapor
RAS	Radio Astronomy Service
RF	radio frequency
RFI	radio frequency interference
RMS	root-mean-square
RNSS	Radionavigation Satellite Service
RR	Radio Regulations (the international treaty governing spectrum use)
SAIR	Synthetic Aperture Interferometric Radiometer
SAR	synthetic aperture radar
SCLP	Snow and Cold Land Processes
SDR	software-defined radio
SETI	Search For Extraterrestrial Intelligence
SFMR	Stepped Frequency Microwave Radiometer
SIRICE	Submillimeter Infrared Radiometer Ice Cloud Experiment
SKA	Square Kilometer Array
SM	soil moisture
SMAP	Soil Moisture Active Passive
SMMR	Scanning Multi-Channel Microwave Radiometer
SMOS	Soil Moisture Ocean Salinity
SNR	signal-to-noise ratio
spfd	spectral power flux density (measured in $Wm^{-2}\,Hz^{-1}$)
SPT	South Pole Telescope
SPTF	Spectrum Policy Task Force
SRS	Space Research Service
SSM/I	Special Sensor Microwave Imager
SSMI/S	Special Sensor Microwave Imager/Sounder
SSM/T	Special Sensor Microwave/Temperature
SSS	sea surface salinity
SST	sea surface temperature
SUBTEL	Subsecretaría de Telecomunicaciones
SSWV	sea surface wind vector
SWE	snow water equivalent

TDM	time-division multiplexing
TDRSS	Tracking and Data Relay Satellite System
TIROS	Television Infrared Observation Satellite
TMI	TRMM Microwave Imager
TOPEX	Topography Experiment
TOVS	TIROS Operational Vertical Sounder
TRMM	Tropical Rainfall Measuring Mission
TTI	total of totals index
UAL	United Airlines
UWB	ultrawideband
VLA	Very Large Array
VLBA	Very Long Baseline Array
VLBI	very long baseline interferometry
WMTS	Wireless Medical Telemetry Service
WP7D	ITU Working Party 7D
WRC	World Radiocommunication Conference

Appendix G

Glossary

altimetry: The measurement of altitude, possibly by using radar.

anisotropic: Having different physical properties along different axes.

anomaly correlation: A common measure of forecast accuracy, with values above 0.6 generally considered to be significant.

antenna main beam: The part of an antenna's radiation pattern (transmit and receive properties) that contains the maximum power and field strength. Synonymous with "beam lobe."

antenna sidelobe: The part of an antenna's radiation pattern (transmit and receive properties) that is not part of the main beam.

array: An interferometric observational scheme that employs multiple linked antennas or dishes to mimic the capabilities of a much larger, single dish.

backhaul: In telecommunications, the intermediate communication link(s) between the end user and the core communications network. For example, for cellular transmissions, the backhaul is the link(s) between the cellular tower and the core communications system.

beam/radiation pattern: The directional dependence of radiation power from the antenna (transmit) or as received by the antenna (receive).

bolometer: An instrument that measures incident electromagnetic radiation.

cross-frequency mitigation: The process of measuring a particular band by splitting it into sub-bands (spectral analysis) and then looking "across the frequencies" for outlier channels. These outliers are presumed to be contaminated with RFI because they have more power in them than would be expected.

direct data assimilation (DDA): A powerful technique developed during the past two decades. DDA optimally uses all available data from satellites, balloons, radars, and surface stations to steer numerical weather prediction (NWP) models. DDA applied to satellite data is known as direct radiance assimilation (DRA) and accounts for most of the improvements in the performance of NWP models in the Southern Hemisphere where other sensors are scarce.

downwelling: Natural radiation that radiates down from the sky.

Environmental Data Record (EDR): Characteristic information regarding an environment that has been observed: once an Earth remote sensing observatory collects incident radiation and it is sent to researchers for processing, the researchers organize the data and interpret them to produce EDRs. EDRs include sea and land wind speed, sea and air temperature, precipitation, sea salinity, and soil moisture. See Table 2.1 in Chapter 2 for a complete list.

Gaussian: A frequency distribution of a variable that exhibits normality and is useful for identifying noise in an instrument.

Global Navigation Satellite System (GLONASS): A radio-based Russian geonavigation satellite constellation operated by the Russian Space Forces that is similar to the United States' Global Positioning System.

interference: The effect of unwanted energy due to one or a combination of emission(s), radiation(s), or induction(s) on reception in a radiocommunications system, manifested by any performance degradation, misinterpretation, or loss of information that could be extracted in the absence of such unwanted energy.

interference mitigation: The process of preempting, identifying, and excising radio or microwave interference from observational data. It can be either unilateral (in the case of excising) or multilateral (in the case of coordination agreements).

interferometry: An observational technique that achieves a large angular resolution by combining the collected radiation from numerous dispersed, linked dishes and examines the resulting interference pattern. The use of interferometry, for which multiple smaller dishes are built, can be more cost-effective than building one enormous dish.

interstellar medium: The physical space between stars that consists of gas, dust, atomic particles, and magnetic fields.

Iridium: A radio-based communications satellite constellation operated by Iridium, Inc., that provides global, handheld, satellite telephone service.

isotropic: Having equal physical properties along different axes.

nowcasting: The practice of forecasting the next 6 hours of weather using observational data. Nowcasting is more precise than forecasting because it has better information on small-scale weather structures.

numerical weather prediction (NWP) models: Computer simulations that forecast future conditions based on information on current weather conditions. A significant limitation of any forecasting model is the reliability and availability of data input.

out-of-band emission: A type of emission that causes interference on a frequency or frequencies immediately outside the necessary bandwidth and that results from the modulation process, but excluding spurious emissions.

Part 15 device: A device that is regulated by Section 15 of Title 47 of the *Code of Federal Regulations* and therefore is not subject to licensing before radiating on an intentional or unintentional basis.

passband/bandpass: The range of frequencies that can pass through a filter.

passive radio/microwave observations: Observations of the natural radio or microwave environment that are made on a receive-only basis—that is, there are no transmissions involved.

polarimetry: The measurement of the polarization of incident radiation that has been reflected and thus provides information on the object off of which the radiation was reflected.

power flux density (pfd): The radio power incident on the antenna, called the flux density or spectral power flux density. Flux density is measured in janskys, where 1 Jy = 10^{-26} Wm^{-2} Hz^{-1}.

quantum transition: The change of an atom or molecule from one quantum state to another.

radio and microwave bands of relevance:
L-band (1-2 GHz)
C-band (4-8 GHz)
X-band (8-10 GHz)
K-band (20-40 GHz)
V-band (50-75 GHz)

radiometry: The study of the measurement of electromagnetic radiation.

radio science: Any scientific endeavor that employs the use of radio or microwave radiation to explore the fundamental characteristics of natural phenomena.

radiosonde: An instrument flown aboard a weather balloon to measure localized, current atmospheric parameters. Radiosonde observations are important inputs to numerical weather prediction models.

Recommendation ITU-R RS.1029: This International Telecommunication Union (ITU) document recommends –134 dBm with a 100 MHz reference bandwidth. The equivalent recommended maximum level using 100 MHz reference bandwidth is –131 dBm.

scatterometry: The measurement of a normalized radar cross section of pulses reflected off a surface. This technique has been particularly useful for the measurement of ocean surface winds.

signal modulation: Varying a periodic waveform.

spectral efficiency: The degree to which a given portion of the spectrum is actually used, compared with the maximum theoretical possible use of that portion.

spectral occupancy: The fraction of time that a transmission can be detected at a given frequency, for a given sensitivity and a given time-frequency resolution.

spectrometer: An instrument that analyzes incident radiation to allow the measurement of individual spectral lines that are characteristic of specific atoms or molecules.

spectroscopy: A technique that analyzes incident radiation to allow the measurement of individual spectral lines that are characteristic of specific atoms or molecules.

spurious emission: Emission that causes interference on a frequency or frequencies that are outside the necessary bandwidth and the level of which may be reduced without affecting the corresponding transmission of information. Spurious emissions include harmonic emissions, parasitic emissions, intermodulation products, and frequency conversion products but exclude out-of-band emissions.

sub-band: A smaller piece of a specified band.

synthetic aperture: A technique that uses the combined collecting area of numerous dispersed dishes to mimic the capabilities of a much larger, single dish.

type-approved device: An emitting device that is approved by the Federal Communications Commission by its type under the *Code of Federal Regulations.*

unwanted emission: Emissions consisting of spurious emissions and out-of-band emissions.

upwelling: Natural radiation that radiates up from Earth.